AF576033

Monitoring and Assessment of Environmental Quality in Coastal Ecosystems

Monitoring and Assessment of Environmental Quality in Coastal Ecosystems

Editor

Sílvia C. Gonçalves

MDPI • Basel • Beijing • Wuhan • Barcelona • Belgrade • Manchester • Tokyo • Cluj • Tianjin

Editor
Sílvia C. Gonçalves
MARE, Marine and
Environmental Sciences Centre
ESTM – IPLeiria, School of
Tourism and Maritime
Technology
Polytechnic of Leiria
Portugal

Editorial Office
MDPI
St. Alban-Anlage 66
4052 Basel, Switzerland

This is a reprint of articles from the Special Issue published online in the open access journal *Environments* (ISSN 2076-3298) (available at: https://www.mdpi.com/journal/environments/special_issues/Environmental_Quality).

For citation purposes, cite each article independently as indicated on the article page online and as indicated below:

LastName, A.A.; LastName, B.B.; LastName, C.C. Article Title. *Journal Name* **Year**, *Volume Number*, Page Range.

ISBN 978-3-0365-2235-7 (Hbk)
ISBN 978-3-0365-2236-4 (PDF)

Cover image courtesy of Sílvia C. Gonçalves

Contents

About the Editor

Sílvia C. Gonçalves is a Senior Researcher at the Marine and Environmental Sciences Centre (MARE) and holds a Ph.D. in Biology (specialization in Ecology) from the University of Coimbra, Portugal. She is presently Coordinating Professor at the School of Tourism and Maritime Technology–Polytechnic of Leiria (Portugal), lecturing mainly in the scientific areas of biology, ecology, physiology, biological diversity, and marine environment. She focuses her research on marine ecology, in the following domains: (1) bio-ecology of crustacean populations on sandy beaches and their potential as ecological indicators of human impacts; (2) structure and function of macrobenthic communities; (3) monitoring and evaluation of environmental quality in coastal ecosystems, especially on sandy beaches and coastal lagoons. Her scientific research activities also include environmental biotechnology (e.g., bioremediation of trace metals by halophyte plants) and aquaculture (e.g., maintenance and reproduction of polychaetes, sea urchins, and fish).

Preface to "Monitoring and Assessment of Environmental Quality in Coastal Ecosystems"

The environmental quality of coastal environments can be assessed and monitored using distinct approaches. The focus can be set on biotic or abiotic compartments, or on both. In this Special Issue, Cagnazzo *et al.* [1] present an application of ground-based infrared thermography technology in coastal environmental monitoring, a rapid method that allowed identifying the presence of beach litter at the Coastal Dunes Regional Natural Park of Ostuni–Fasano in Apulia (Southern Italy). This research enabled the development and validation of an empirical equation to calculate the sandy-soil surface temperature by knowing only the air temperature, providing an effective tool to detect the presence of anthropogenic polluting material (plastic, glass and rubber) on the studied sandy shore. Also focused on the abiotic compartment, Thomas *et al.* [2] assessed the ability of the Takagi–Sugeno (TS) fuzzy modelling approach with fuzzy c-means (FCM) clustering to obtain spatial predictions of lead concentrations in a marine sediment geochemical dataset. The main aim of the study was to test if fuzzy modelling could still produce a suitable pollutant distribution map using fewer sampling points, potentially reducing the cost associated with new remediation projects. The results demonstrated that TS fuzzy modeling using FCM clustering and constant-width Gaussian-shaped membership functions, did not outperform the inverse distance weighting (IDW) and the ordinary kriging (OK) methods. Thus, this method appears to be unsuitable for use in contaminants remediation projects with sparsely distributed geospatial sampling points.

When biotic compartments are included in the assessment and monitoring approaches to evaluate the environmental quality of coastal environments, the tools used can range between ecological levels of organization, from individuals to the ecosystem. Distinct types of biological responses can also be assessed (*e.g.* bioecology, ecotoxicology, physiology and behavior).

From the ecotoxicological perspective, Cocci *et al.* [3] used a combined in silico and in vitro approach to evaluate the impact of two aquatic emerging pollutants capable of acting as endocrine disrupting chemicals (EDCs), perfluorononanoic acid (PFNA) and enalapril (ENA), on the grey mullet (*Mugil cephalus*) hepatic estrogen signaling pathway. These EDCs tend to bioaccumulate in aquatic organisms at concentrations that may cause reproductive toxicity. Their results showed that ENA has a weak agonist activity on estrogen receptors α, whereas PFNA showed moderate-to-high agonist binding to both tested estrogen receptors (α and β). Hepatocytes' incubation for 48 h to PFNA caused a concentration-dependent upregulation of the estrogen receptors and vitellogenin gene expression profiles, whereas only a small increase was observed in estrogen receptors' mRNA levels for the highest ENA concentration, suggesting a structure–activity relationship between hepatic estrogen receptors and these pollutants.

Focusing on a different type of marine pollutant, Biswas *et al.* [4] analyzed the radioactivity and trace metal levels in six marine fish and four crustacean species, all edible, of the northern coastal belt of the Bay of Bengal (Bangladesh). To assess the environmental quality of the region concerning those pollutants and the health risks posed to humans from the consumption of those fish and crustaceans, several radiological and health-hazard parameters were calculated. The research findings demonstrate an increase in the pollution due to radioactivity and trace metals in this coastal belt of the Bay of Bengal, caused by an increase in human activities in the region. It was also found that consuming the studied species from the Bay of Bengal may have adverse health impacts if consumption and/or the pollution sources are not controlled.

Suitable assessment and monitoring programs of environmental quality at the ecosystem level require a profound knowledge of the services rendered by the ecosystem under analysis, as well as a recognition of its importance across multiple dimensions (*e.g.* ecological, socio-cultural and economic). Determining ecosystem service values (ESVs) is extremely relevant, namely to better communicate the importance of protecting ecologically functional ecosystems and biodiversity to decision-makers. In this Special Issue, Magalhães Filho *et al.* [5] used the Ecosystem Service Valuation Database (ESVD) to estimate meta-regression functions for provisioning, regulating and maintenance, and cultural ecosystem services across 12 biomes at the global scale. The research findings demonstrate that among the biomes with the highest ESVs are coral reefs, inland wetlands, and coastal wetlands, which, amongst other characteristics, are transitional, aquatic-terrestrial biomes that are scarce and provide a wide diversity of services. The authors concluded that when considering the characteristics of the study area under analysis, the inclusion of explanatory variables such as income, population density, and protection status, it is possible to determine the value of ecosystem services with higher accuracy.

Sílvia C. Gonçalves
Editor

 environments 

Article

Autumnal Beach Litter Identification by Mean of Using Ground-Based IR Thermography

Cosimo Cagnazzo [1,*], Ettore Potente [1], Hervé Regnauld [2], Sabino Rosato [3] and Giuseppe Mastronuzzi [1]

[1] Department of Earth and Geo-Environmental Science, University of Bari, 70121 Bari, Italy; ettore.potente@uniba.it (E.P.); giuseppe.mastronuzzi@uniba.it (G.M.)
[2] Littoral, Environnement, Télédétection, Géomatique, University of Rennes 2, 35043 Rennes, France; herve.regnauld@univ-rennes2.fr
[3] Serveco Srl, 74020 Montemesola, Italy; s.rosato@serveco.it
* Correspondence: cosimo.cagnazzo@uniba.it

Abstract: The progress of scientific research and technological innovation are contributing to an increase in the use of rapid systems for monitoring and identifying geo-environmental processes related to natural and/or anthropogenic activities. The aim of this study is identifying autumnal beach litter using ground-based IR thermography. Starting from quarterly autumn monitoring data of air temperature and sandy soil surface temperature, an empirical equation between the two environmental matrices (air and sandy soil) is obtained. This will allow the calculation of the sandy soil surface temperature knowing only the air temperature. Therefore, it will be possible to know in advance the thermal response of the sandy soil, thus creating a thermal blank of the beach. Using an IR thermal camera, it is possible for a quicker identification of thermal anomalies of the coastal area potentially connected to the presence of pollution due to the anthropogenic origin (particularly plastic material). The test area is located in the area of the Coastal Dunes Regional Natural Park of Ostuni–Fasano in Apulia (southern Italy).

Keywords: beach litter; infrared thermography; UAV; UGV; environmental monitoring; coastal pollution

Citation: Cagnazzo, C.; Potente, E.; Regnauld, H.; Rosato, S.; Mastronuzzi, G. Autumnal Beach Litter Identification by Mean of Using Ground-Based IR Thermography. *Environments* **2021**, *8*, 37. https://doi.org/10.3390/environments8050037

Academic Editor: Sílvia C. Gonçalves

Received: 22 March 2021
Accepted: 22 April 2021
Published: 24 April 2021

1. Introduction

Among the atmospheric and non-atmospheric parameters that influence the properties of the climate system, temperature is a very relevant parameter in all the chemical, physical and biological processes that affect the soil formation and its persistence in a natural environment. Considering the same source of thermal radiation, each material is heated differently according to its chemical-physical characteristics. Scientific research and technological innovation have developed new methodologies (IR thermography) and new investigation tools (thermal cameras) based on the use of temperature for the identification of different materials, such as sandy soil. IR thermography allows for detecting and quantifying the infrared energy emitted by any object above absolute zero temperature (−273.14 °C).

Since every object characterized by a temperature above absolute zero emits thermal energy, it can be identified through infrared thermography. Thermal cameras allow for quick and remote collection of a large amount of data.

Over the last few years, the use of thermography has become widespread in various fields such as: research and development, quality and process control, healthcare, construction, industry and the mechanical field [1–4]. In the environmental monitoring field, thermography had a remarkable development, especially in wildfire detection [5], while is only a first approach for flora and vegetation habitat monitoring, and sometimes also for wildfauna, which necessarily require observations and confirmations directly in the field.

Moreover, the growing anthropogenic impact on the natural coastal environment is causing a significant increase in coastal pollution [6] and spills of chemical products in the marine environment [7]. The need to identify the areas affected by such events is affirming the use of thermography as a new technique, complementary to other traditional methodologies [8–10].

IR thermal cameras are also installed on aerial (UAV) and terrestrial (UGV) platforms, resulting in effective equipment in terms of time efficiency and timeliness [11–16].

The aim of this study is identifying autumnal beach litter using ground-based IR thermography. Starting from statistical analysis results of monitoring data collected by two air temperature and sandy-soil temperature sensors, installed in the Coastal Dunes Regional Natural Park (Ostuni–Fasano, Italy) (Figure 1), an empirical relation between the environmental matrices (air and sandy soil) will be found. This will allow to estimate the sandy-soil surface temperature knowing only the air temperature.

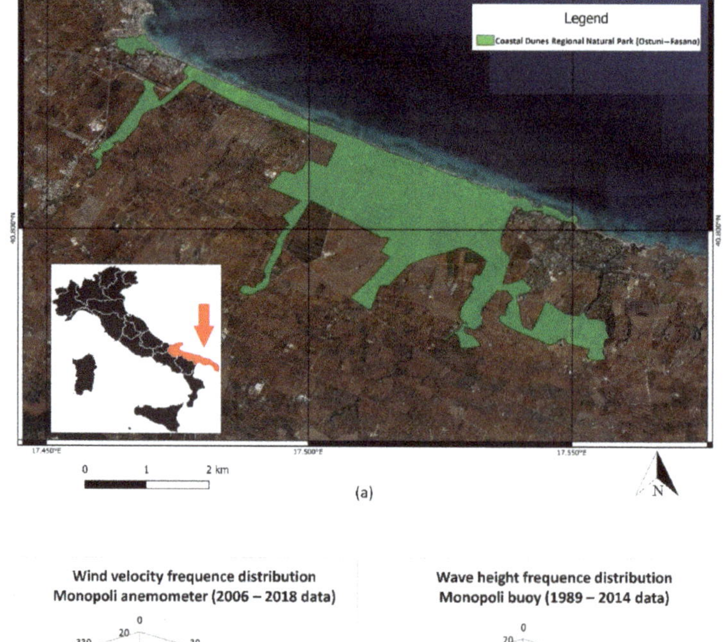

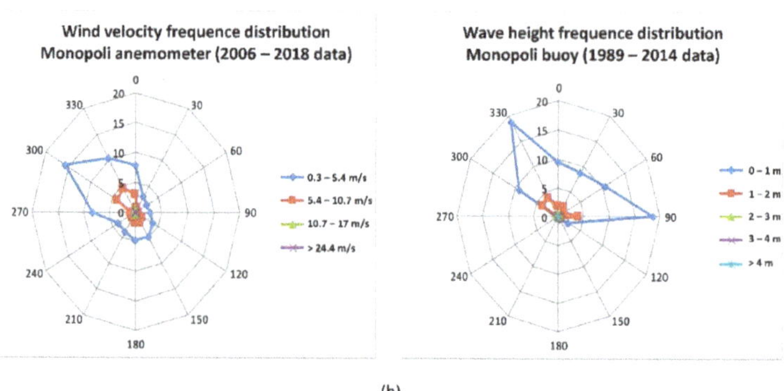

Figure 1. (**a**) Study area localization in the Coastal Dunes Regional Natural Park (Apulia, Italy); (**b**) Polar plots of the wind speed distribution and wave height distribution.

By estimating the sandy-soil surface temperature it will be possible to rapidly detect any thermal anomalies along the sandy coast, such as anomalies deriving from plastic pollution, by using thermal cameras installed on aerial or terrestrial platforms. The results of this study will help with the optimization of infrared thermography applications in the coastal environmental monitoring.

1.1. Ecological Setting

The Coastal Dunes Regional Natural Park extends from the coast towards inland agricultural areas, occupied by centuries-old olive groves. It includes the Site of Community Importance (SIC) "Litorale Brindisi", included in the European network "Natura 2000". In the Park are many species of flora that are well preserved, such as the psammophilous and Cakiletum maritimae [17–19]. The system erosion and the relative vegetation are not caused by the wind but are mainly due to people leveling the sand in the summer season, especially in the investigated site and in general on the whole psammophilous coastal area that extends from the site study and far as Brindisi.

The Park was established with the aim of conserving and recovering the habitats and animal and plant species indicated in Community Directives 79/409/EEC and 92/43/EEC.

1.2. Geological and Geomorphological Setting

The area covered by this study is located on the Apulian Adriatic coast, in the Coastal Dunes Park, stretched for 6 km from Torre Canne to Torre San Leonardo. The area hosts a coastal mobile system characterized by the presence of several lakes and ponds and by a polyphasic dune belt parallel to the coast that reaches an altitude of about 17 m (Figure 2).

Figure 2. View of the study area (19 July 2018). The characteristics of the beach change quickly and dramatically depending on the storm surge.

The deposits consist of medium sand characterized by carbonates, quartz and other minerals in very small percentages (pyroxene and feldspar); there are also rare fragments of siliciclastic rocks and material of anthropogenic origin [20–22].

1.3. Climatological Setting

The average annual temperature of the study area ranges between 15 and 16 °C according to 1971–2000 monitoring data [23].

The whole coastal sector is exposed to a wind regime characterized by winds coming from the north-western quadrants (Figure 1), therefore storm surges are very frequent in winter and also cause the stranding of a large amount of natural and anthropogenic material. The main direction of the longshore transport is NW-SE [24] (Figure 1).

2. Materials and Methods

In order to create a thermal blank of the beach and consequently identify the thermal polluting anomalies (attributable to the beach litter) with the thermal camera (FLIR C3), two sensors (Elitech RC-4) equipped with a data logger were installed (Figures 3 and 4). Sandy soil surface temperature data and air temperature data were collected. Soil moisture affects soil temperature just as air humidity affects air temperature. Furthermore, the humidity of the air could be influenced by other atmospheric parameters (rain, wind, atmospheric pressure) and the humidity of the sandy soil by other parameters (vegetation, storm surges, rain). Since the thermal imager acquires only the temperature data, the experimental monitoring activity concerned only the temperature data (air and soil). This was done to make the thermographic technique a methodology to support other new methodologies (spectral sensors, artificial intelligence algorithms) in emergency coastal pollution conditions where it is necessary to be quick in identification and mapping.

Figure 3. Sensor localization; and IR thermography survey area.

(**a**) (**b**)

Figure 4. Air temperature sensor (**a**); sandy-soil surface temperature sensor (**b**). The pictures clearly show the presence of vegetation typical of the dune environment (Cakiletum, Ammophiletum and Agropyretum).

The first thermal sensor inserted in a meteorological screen (solar radiation protection), was installed 2 m from the ground to measure air temperature (Figure 4a) as recommended by WMO (World Meteorological Organization) guidelines. The sensor is able to measure temperatures ranging from −40 °C to +80 °C with a 0.1 °C resolution and a 0.5 °C accuracy. The Elitech RC-4 sensor with probe was used for the sandy soil surface temperature. It has the same technical specs, and the data logger was positioned on the dune into a waterproof box, protected from bad weather and anthropic interference, and the probe (connected to the data logger) directly in contact with the sandy soil surface (Figure 4b).

Both sensors were set to a sampling rate of 5 min, the monitoring started on 7 October 2018 and ended on 15 December 2018. The monitoring period was chosen to avoid anthropogenic disturbances related to tourism, which could compromise the sensors installed. The collected data were downloaded and exported in .txt format, in a dataset gathering about 20,000 thermal data for each sensor. The statistical analysis was carried out using Microsoft Excel 2019 and R 4.0.3 software package. Daily average air and sandy-soil surface temperature was calculated, as well as Pearson's correlation coefficient. This coefficient is the test that measures the statistical relationship between two continuous variables, giving information about the correlation and the direction of the relationship, and it ranges between −1 (strong negative linear relationship) and +1 (strong positive linear relationship). A null coefficient means there is no linear relationship between the variables.

Moreover, by analysing the scatter plot, the line of best fit and the empirical equation were obtained. The coefficient of determination R^2 was calculated to understand the accuracy of the regression model used to make predictions. The coefficient ranges between 0 and 1 and it is a statistical measure of how close the data are to the fitted regression line. Finally, the RMSE (Root Mean Squared Error) was calculated, and the daily average temperature values observed were compared with the ones obtained from the prediction model.

Six thermal images 80 × 60 (4800 pixels) on the ground were collected in the study area, using an FLIR C3 thermal imaging camera (Figure 3). Its sensor can detect and measure temperatures between −10 °C and +150 °C in a spectral band ranging from 7.5 to 14 μm, to a resolution of 0.1 °C and an accuracy of ±2 °C. The images were collected on 30 October 2018, twice in the morning (8:00 AM UTC and 9:00 AM UTC) and twice in the afternoon (12:00 PM UTC and 2:30 PM UTC); and on 31 October 31 in the morning (5:30 AM UTC) and in the afternoon (01:00 PM UTC). The observed air temperature was:

- 16.5 °C ± 0.5 and 17.1 °C ± 0.5 (on 30 October 2018, respectively, at 08:00 AM UTC and at 09:00 AM UTC);
- 24.8 °C ± 0.5 and 19.5 °C ± 0.5 (on 30 October 2018, respectively, at 12:00 PM UTC and at 02:30 PM UTC);
- 14.4 °C ± 0.5 and 23.9 °C ± 0.5 (on 31 October 2018, respectively, at 05:30 AM UTC and at 01:00 PM UTC).

The thermal images were processed with FLIR Tools software, setting the sand emissivity (ε = 0.90). For each radiometric thermal acquisition, knowing the air temperature, the empirical equation was applied in order to obtain a prediction of the sandy soil thermal range, taking into account the instrumental error.

Finally, for each thermal image processed, the GIS software created 10 random points. Later the same points were imported into the corresponding RGB images to understand what they really identified. It was assessed that the degree of confidence of the results through the calculation of the kappa coefficient [25,26] allowed for the assessment of the degree of confidence of the results on a morning and afternoon scale. The kappa index measures the agreement between different assessments for the classification of the same object. Kappa index is defined as:

$$K = (po - pe)/(1 - pe) \tag{1}$$

po = observed agreement; pe = hypothetical probability of agreement.

The calculation of the index is carried out with the construction of a matrix called the confusion matrix.

This index ranges from 0 to 1 and expresses the correlation between the homologues points in the IRT and RGB image. K = 1 indicates perfect agreement while K = 0 indicates absence of agreement. Generally, a value of the index k > 0.75 indicates a good agreement.

3. Results and Discussion

Daily average air temperature and sandy-soil surface temperature are shown in Figure 5, while Table 1 shows the maximum and minimum values of the daily average air temperature and sandy soil surface temperature. Figure 5 shows that the sandy-soil surface temperature is always lower than the air temperature. This happens because in the autumn the surface of the sandy soil, in this study area, is always wet.

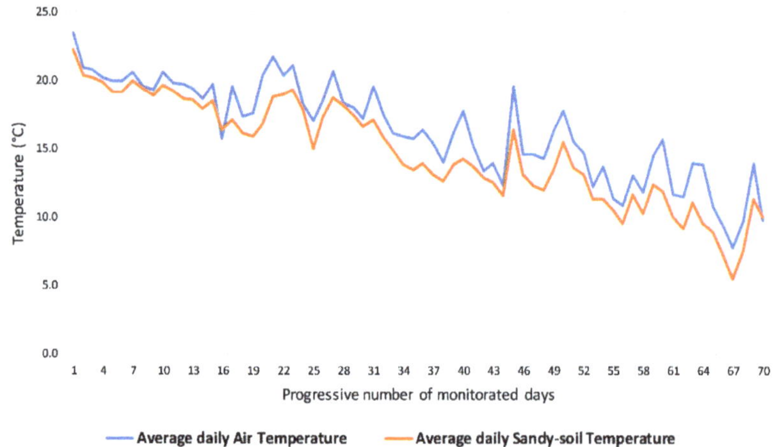

Figure 5. Daily average air and sandy soil surface temperature during the monitoring period.

Table 1. Extreme values of average temperature.

	Air	Sandy-Soil Surface
Maximum daily average temperature (°C)	23.5	22.2
Minimum daily average temperature (°C)	7.8	5.5

The Pearson correlation coefficient shows a strong positive correlation (0.97) between average daily air temperature and average daily sandy-soil surface temperature soil. The best model for the study area, approximating the empirical data, was calculated from the scatter plot (Figure 6) and it results in a non-linear regression, a power function. In fact, in the interpolation curve, a slight upwards concavity can be observed.

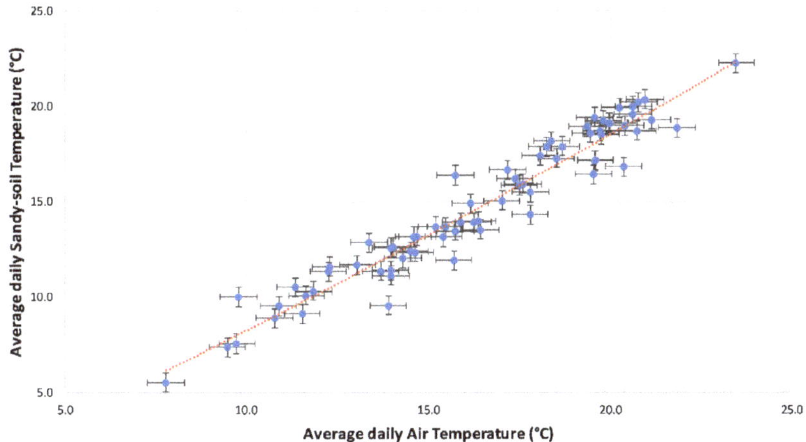

Figure 6. Scatter plot between average air temperature and sandy soil surface temperature.

The resulting model is represented by the equation:

$$Ts = 0.5572 \times Ta^{1.1701} \quad (2)$$

Ts = sandy-soil surface temperature; Ta = air temperature.

The coefficient of determination $R^2 = 0.94$ has validated the statistical model. The empirical equation obtained was used to predict the daily average sandy-soil surface temperature knowing the air temperature. Therefore, the results were compared with the temperature measured on site (Figure 7).

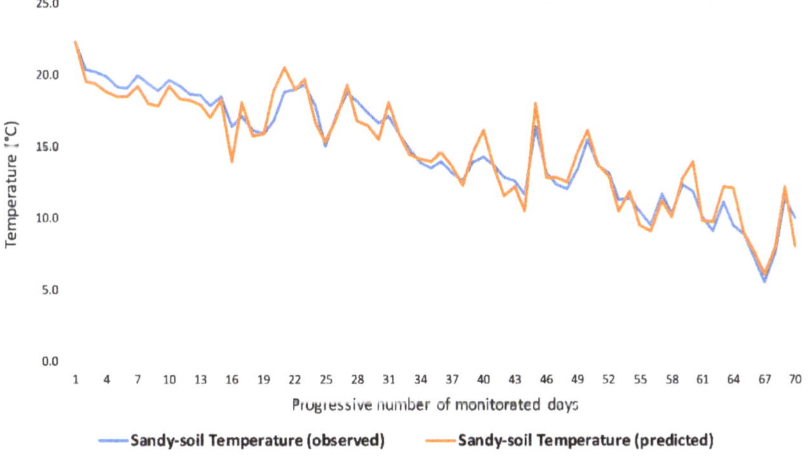

Figure 7. Comparison between measured and predicted sandy soil surface temperature during the same monitoring period.

RMSE has been obtained through statistical processing. It is a statistical indicator that quantifies the deviation between the observed and simulated data. The value of RMSE stood at 0.96. The very low value confirms the accuracy of the empirical equation obtained. For statistical completeness, a linear regression of the data was also performed. Although the value of $R^2 = 0.94$ was equal to the non-linear regression, the value of the RMSE (0.97) was slightly higher than the non-linear regression. This statistical difference influenced the choice of the regression typology to obtain the empirical equation.

Table 2 shows the air temperature values obtained from the morning and afternoon thermal imaging survey. Sandy-soil surface temperature values was obtained by using these values in the Equation (2). Moreover, taking into account the thermal camera accuracy, the thermal range of the sandy soil surface is shown.

Table 2. Temperature data (30–31 October 2018).

	Air Temp. (°C) Observed	Sandy Soil Surface Temp. (°C) Predicted by Applying (2)	Potential Thermal Range of Sandy Soil Surface (for Each IR Image) Considering the Previous Column and the Instrumental Accuracy (±2 °C)
08:00 AM UTC	16.5	14.8	12.8–16.8
09:00 AM UTC	17.1	15.5	13.5–17.5
12:00 PM UTC	24.8	20.5	18.5–22.5
02:30 PM UTC	19.5	18	16–20
05:30 AM UTC	14.4	12.6	10.6–14.6
01:00 PM UTC	23.9	22.8	20.8–24.8

Figure 8 shows the thermal images collected in the morning of 30 October 2018 and 31 October 2018, processed using FLIR Tools. Table 3 shows the correspondence of the material detected in the thermal image (prediction) compared with the RGB image (truth) in the randomly selected points.

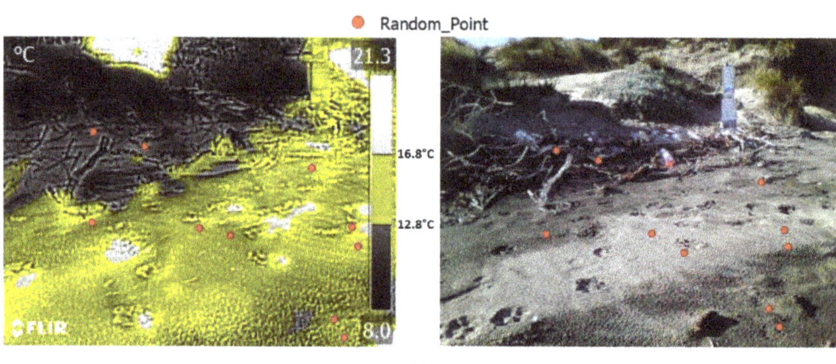

(a)

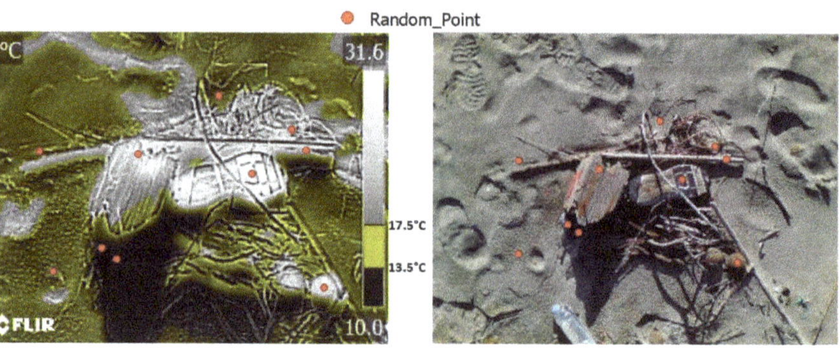

(b)

Figure 8. *Cont.*

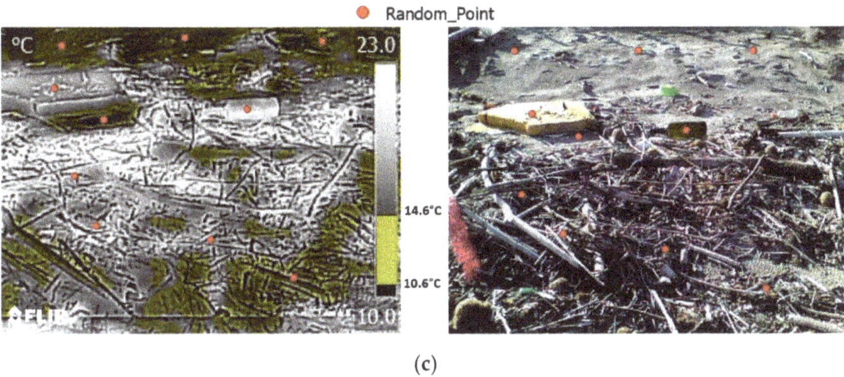

(c)

Figure 8. Morning acquisition: thermal image (on the left) and RGB image (on the right). Randomly selected points (in red). The yellow pixels in the thermal image indicate the sandy soil surface identified by the application of the empirical Equation (2). Plastic bottles can be seen in (**a**), rubber soles in (**b**), glass bottles and expanded polystyrene in (**c**).

Table 3. Correspondence between the material detected in the thermal image and the RGB image in the selected points (morning).

		Material in the RGB Image (Truth)		
		Sandy Soil	**Natural and/or Anthropic Anomaly**	**Total**
Material detected in the thermal image (prediction)	Sandy soil	15	1	16
	Natural and/or anthropic anomaly	0	14	14
	Total	15	15	30

The kappa coefficient value was calculated and resulted to be equal to 0.93. Figure 9 shows the thermal images collected in the afternoon of 30 October 2018 and 31 October 2018 and processed. Table 4 shows the correspondence between of the material detected in the thermal image (prediction) compared with the RGB image (truth) in 30 random points.

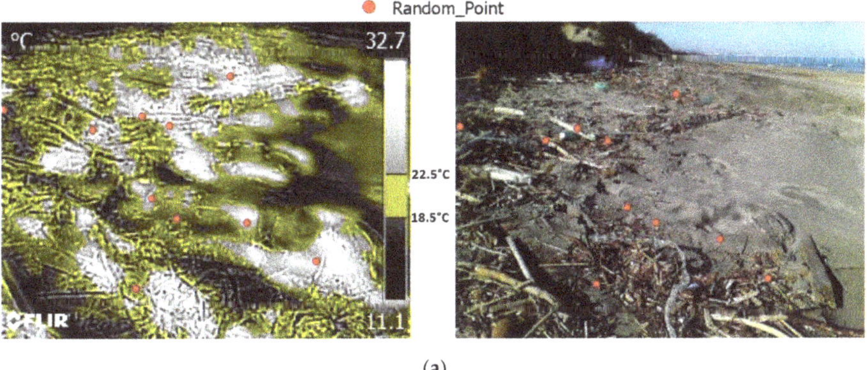

(a)

Figure 9. *Cont.*

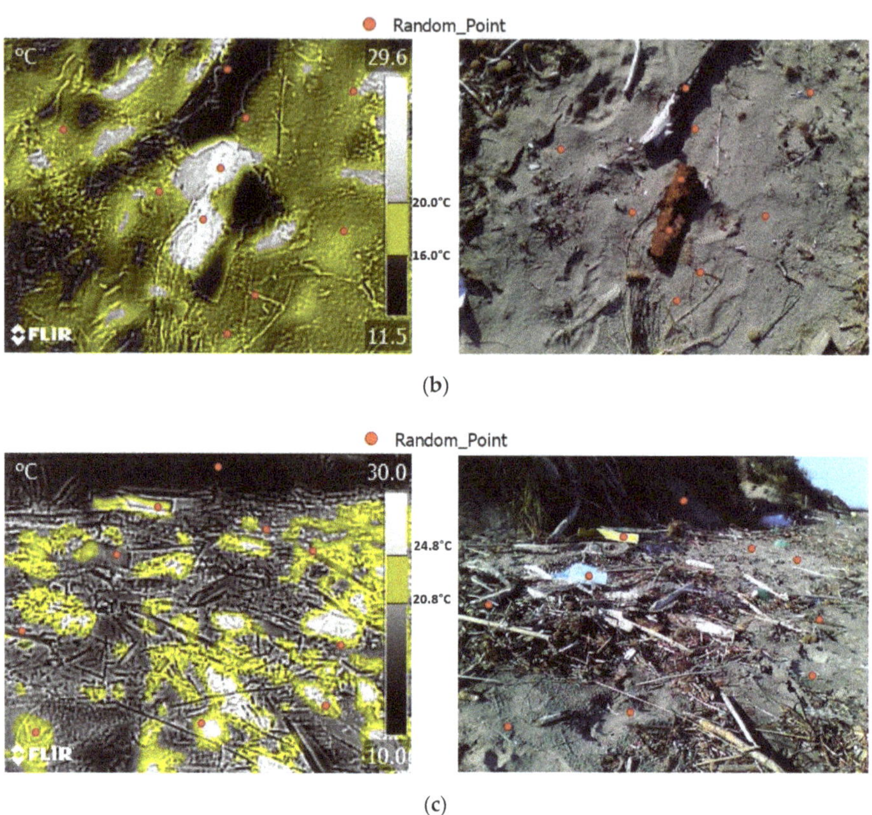

Figure 9. Afternoon acquisition: thermal image (on the left) and RGB image (on the right). Randomly selected points (in red). The yellow pixels in the thermal image represent the sandy soil surface identified by the application of the empirical Equation (2). Plastic bottles and bags can be seen in (**a**,**c**) and rusty bottles in (**b**).

Table 4. Correspondence between the material detected in the thermal image and the RGB image in the selected points (afternoon).

		Material in the RGB Image (Truth)		
		Sandy Soil	**Natural and/or Anthropic Anomaly**	**Total**
Material detected in the thermal image (prediction)	Sandy soil	13	1	14
	Natural and/or anthropic anomaly	3	13	16
	Total	16	14	30

The kappa coefficient value was calculated and resulted to be equal to 0.74. The results obtained from empirical equation application in the thermal images, the comparison between the material detected in the thermal images and the values of the kappa coefficient, allowed for validation of the thermal methodology used, which was developed to identify sandy soil on the coast using IR thermography, distinguishing it from thermal anomalies which can be due to the presence of different materials and polluting objects and material of anthropogenic origin such as plastic. The methodology described may have problems with shaded areas, because they are significantly cooler than the predicted soil temperature value. However, using UAV systems with larger scale acquisitions, the shaded areas would have less weight in the acquired thermograms than the really radiated beach areas.

4. Conclusions

This study represents an application of the IR thermography technology in the field of coastal environmental monitoring. The 70-day monitoring of the air and sandy soil surface temperature by means of two thermal sensors, carried out in the Coastal Dunes Regional Natural Park (Ostuni–Fasano), and the statistical analysis of a considerable amount of thermal data, allowed the development of an empirical equation used for calculating the sandy-soil surface temperature by knowing only the air temperature. The existence of a strong correlation between the two variables, the high value of the coefficient of determination R^2 of the model and a low value of the RMSE value, confirmed the good quality of the empirical equation used.

Moreover, the tests carried out applying the equation to the IR thermal survey and the value of the kappa coefficient calculated, validated this methodology, making it an effective tool for anthropogenic polluting material detection (plastic, glass, rubber) on the sandy coast in the study area.

The results obtained can be used to rapidly process thermal images deriving from surveys carried out with sensors installed on UAV and UGV, quickly detecting the presence of anomalies due to potentially polluting objects (like plastic, glass, etc) or other pollutant materials.

Further coastal environmental geology studies will be carried out in the future. In particular, remote multispectral methodologies will be tested in this study area for a more precise identification of pollutants.

Author Contributions: Conceptualization (C.C.), Data curation (C.C.), Formal analysis (C.C.; E.P. and S.R.), Founding acquisition (S.R. and G.M.), Investigation (C.C. and E.P.), Methodology (C.C.), Project administration (G.M.), Resources (S.R. and G.M.), Software (C.C.), Supervision (H.R.; S.R. and G.M.), Validation (C.C.; E.P.; S.R. and G.M.), Visualization (C.C. and E.P.), Writing—original draft (C.C. and E.P.), Writing—review and editing (C.C.; E.P.; S.R. and G.M.). All authors have read and agreed to the published version of the manuscript.

Funding: Contributo Protezione Civile Regione Puglia–Progetto I-STORMS; Scientific Supervisor: Giuseppe Mastronuzzi.

Institutional Review Board Statement: Not applicable.

Informed Consent Statement: Not applicable.

Data Availability Statement: The data presented in this study are available on request from the corresponding author.

Acknowledgments: The authors would like to thank the reviewers for their contributions and constructive comments for the improvement of the manuscript and Regional Natural Park of the Coastal Dunes for their kind support. Work carried out as part of the research project "SIAT—Integrated UAV–UTV system with remote control for the pollutants dispersion assessment in the coastal environment for remediation" related to the XXXIII PON Cycle of the Ph.D. in Geosciences of the University of Bari, under the responsibility of Massimo Moretti.

Conflicts of Interest: The authors declare no conflict of interest.

References

1. Bianchi, F.; Pisello, A.L.; Baldinelli, G.; Asdrubali, F. Infrared Thermography Assessment of Thermal Bridges in Building Envelope: Experimental Validation in a Test Room Setup. *Sustainability* **2014**, *6*, 7107–7120. [CrossRef]
2. Costanzo, A.; Minasi, M.; Casula, G.; Musacchio, M.; Buongiorno, M.F. Combined Use of Terrestrial Laser Scanning and IR Thermography Applied to a Historical Building. *Sensors* **2015**, *15*, 194–213. [CrossRef] [PubMed]
3. Garduño-Ramón, M.A.; Vega-Mancilla, S.G.; Morales-Henández, L.A.; Osornio-Rios, R.A. Supportive Noninvasive Tool for the Diagnosis of Breast Cancer Using a Thermographic Camera as Sensor. *Sensors* **2017**, *17*, 497. [CrossRef] [PubMed]
4. Kim, J. Non-Destructive Characterization of Railway Materials and Components with Infrared Thermography Technique. *Materials* **2019**, *12*, 4077. [CrossRef] [PubMed]
5. Aranda, J.M.; Melendez, J.; De Castro, A.J.; Lopez, F. Forest fire studies by medium infrared and thermal infrared thermography. In Proceedings of the Thermosense XXIII, Orlando, FL, USA, 23 March 2001; Volume 4360.

6. Cagnazzo, C.; Potente, E.; Rosato, S.; Mastronuzzi, G. Geostatistics and Structure from Motion Techniques for Coastal Pollution Assessment along the Policoro Coast (Southern Italy). *Geosciences* **2020**, *10*, 28. [CrossRef]
7. Cagnazzo, C.; Potente, E.; Rosato, S.; Mastronuzzi, G. Numerical modeling for coastal environmental monitoring: Oil spills simulation along the lower Adriatic coasts. *Rend. Online Soc. Geol. It.* **2020**, *52*, 19–24. [CrossRef]
8. Manfreda, S.; McCabe, M.F.; Miller, P.E.; Lucas, R.; Pajuelo Madrigal, V.; Mallinis, G.; Ben Dor, E.; Helman, D.; Estes, L.; Ciraolo, G.; et al. On the Use of Unmanned Aerial Systems for Environmental Monitoring. *Remote Sens.* **2018**, *10*, 641. [CrossRef]
9. Stuart, M.B.; McGonigle, A.J.S.; Willmott, J.R. Hyperspectral Imaging in Environmental Monitoring: A Review of Recent Developments and Technological Advances in Compact Field Deployable Systems. *Sensors* **2019**, *19*, 3071. [CrossRef] [PubMed]
10. Zanutta, A.; Lambertini, A.; Vittuari, L. UAV Photogrammetry and Ground Surveys as a Mapping Tool for Quickly Monitoring Shoreline and Beach Changes. *J. Mar. Sci. Eng.* **2020**, *8*, 52. [CrossRef]
11. Casana, J.; Kantner, J.; Wiewel, A.; Cothren, J. Archaeological aerial thermography: A case study at the Chaco-era Blue J community, New Mexico. *J. Archaeol. Sci.* **2014**, *45*, 207–219. [CrossRef]
12. Lega, M.; Kosmatka, J.; Ferrara, C.; Russo, F.; Napoli, R.M.A.; Persechino, G. Using Advanced Aerial Platforms and Infrared Thermography to Track Environmental Contamination. *Environ. Forensics J.* **2012**, *13*, 332–338. [CrossRef]
13. Lagüela, S.; Díaz-Vilariño, L.; Roca, D.; Lorenzo, H. Aerial thermography from low-cost UAV for the generation of thermographic digital terrain models. *Opto-Electron. Rev.* **2015**, *23*, 78–84. [CrossRef]
14. Ficapal, A.; Mutis, I. Framework for the Detection, Diagnosis, and Evaluation of Thermal Bridges Using Infrared Thermography and Unmanned Aerial Vehicles. *Buildings* **2019**, *9*, 179. [CrossRef]
15. Savelyev, A.; Sugumaran, R. Surface Temperature Mapping of the University of Northern Iowa Campus Using High Resolution Thermal Infrared Aerial Imageries. *Sensors* **2008**, *8*, 5055–5068. [CrossRef] [PubMed]
16. Szrek, J.; Wodecki, J.; Błażej, R.; Zimroz, R. An Inspection Robot for Belt Conveyor Maintenance in Underground Mine—Infrared Thermography for Overheated Idlers Detection. *Appl. Sci.* **2020**, *10*, 4984. [CrossRef]
17. Ciola, G.; Maiorano, F.; Massari, M.A. Il Parco Naturale Regionale delle Dune Costiere da Torre Canne a Torre San Leonardo: Il valore della biodiversità per ricostruire comunità solidali. Territori e comunità. In *Le sfide dell'autogoverno comunitario. Atti dei Laboratori del VI Convegno della Società dei Territorialisti Castel del Monte (BA)*; Gisotti, M.R., Rossi, M., Eds.; Sdt Edizioni: Bari, Italy, 2018.
18. Perrino, E.V.; Ladisa, G.; Calabrese, G. Flora and plant genetic resources of ancient olive groves of Apulia (southern Italy). *Genet. Resour. Crop Evol.* **2014**, *61*, 23–53. [CrossRef]
19. Tomaselli, V.; Tenerelli, P.; Sciandrello, S. Mapping and quantifying habitat fragmentation in small coastal areas: A case study of three protected wetlands in Apulia (Italy). *Monit. Assess.* **2012**, *184*, 693–713. [CrossRef]
20. Mastronuzzi, G.; Palmentola, G.; Sanso, P. Evoluzione morfologica della fascia costiera di Torre Canne (Puglia adriatica). *Studi Costieri* **2001**, *4*, 19–31.
21. Mastronuzzi, G.; Sanso, P. Holocene coastal dune development and environmental changes in Apulia (Southern Italy). *Sediment. Geol.* **2002**, *150*, 139–152. [CrossRef]
22. Moretti, M.; Tropeano, M.; Van Loon, A.J.; Acquafredda, P.; Baldacconi, R.; Festa, V.; Lisco, S.; Mastronuzzi, G.; Moretti, V.; Scotti, R. Texture and composition of the Rosa Marina beach sands (Adriatic coast, southern Italy): A sedimentological/ecological approach. *Geologos* **2016**, *22*, 87–103. [CrossRef]
23. Marsico, A.; Caldara, M.; Capolongo, D.; Pennetta, L. Climatic characteristics of middle-southern Apulia (southern Italy). *J. Maps* **2007**, *3*, 342–348. [CrossRef]
24. Mastronuzzi, G.; Palmentola, G.; Sansò, P. Lineamenti e dinamica della costa pugliese. *Studi Costieri* **2002**, *5*, 9–22.
25. Cohen, J. A coefficient of agreement for nominal scales. *Educ. Psychol. Meas.* **1960**, *20*, 37–46. [CrossRef]
26. Cohen, J. Weighed kappa: Nominal scale agreement with provision for scaled disagreement or partial credit. *Psychol. Bulletin.* **1968**, *70*, 213–220. [CrossRef] [PubMed]

Communication

An Investigation of Takagi-Sugeno Fuzzy Modeling for Spatial Prediction with Sparsely Distributed Geospatial Data

Robert Thomas [1], Usman T. Khan [2], Caterina Valeo [1,*] and Fatima Talebzadeh [1]

[1] Department of Mechanical Engineering, University of Victoria, Victoria, BC V8W 2Y2, Canada; thomas.rwc@gmail.com (R.T.); talebzadeh@uvic.ca (F.T.)

[2] Department of Civil Engineering, Lassonde School of Engineering, Toronto, ON M3J 1P3, Canada; usman.khan@lassonde.yorku.ca

* Correspondence: valeo@uvic.ca

Abstract: Fuzzy set theory has shown potential for reducing uncertainty as a result of data sparsity and also provides advantages for quantifying gradational changes like those of pollutant concentrations through fuzzy clustering based approaches. The ability to lower the sampling frequency and perform laboratory analyses on fewer samples, yet still produce an adequate pollutant distribution map, would reduce the initial cost of new remediation projects. To assess the ability of fuzzy modeling to make spatial predictions using fewer sample points, its predictive ability was compared with the ordinary kriging (OK) and inverse distance weighting (IDW) methods under increasingly sparse data conditions. This research used a Takagi–Sugeno (TS) fuzzy modelling approach with fuzzy c-means (FCM) clustering to make spatial predictions of the lead concentrations in soil. The performance of the TS model was very dependent on the number of outliers in the respective validation set. For modeling under sparse data conditions, the TS fuzzy modeling approach using FCM clustering and constant width Gaussian shaped membership functions did not show any advantages over IDW and OK for the type of data tested. Therefore, it was not possible to speculate on a possible reduction in sampling frequency for delineating the extent of contamination for new remediation projects.

Keywords: fuzzy modelling; marine sediment; Takagi–Sugeno; ordinary kriging (OK); inverse distance weighting (IDW); spatial predictions

Citation: Thomas, R.; Khan, U.T.; Valeo, C.; Talebzadeh, F. An Investigation of Takagi-Sugeno Fuzzy Modeling for Spatial Prediction with Sparsely Distributed Geospatial Data. *Environments* **2021**, *8*, 50. https://doi.org/10.3390/environments8060050

Academic Editor: Sílvia C. Gonçalves

Received: 23 April 2021
Accepted: 28 May 2021
Published: 29 May 2021

1. Introduction

The release of pollutants into the natural environment has been a problem of global concern since the beginning of the industrial revolution. Highly toxic persistent environmental pollutants often occur in marine harbour sediments as a result of industrial practices around the world and pose a significant risk to human health [1]. The major contributor to exposure of humans to contaminated sediment is through the ingestion of contaminated food as a result of bioaccumulation through the food chain. Chronic exposure to contamination can lead to infertility, birth defects, impaired child development, diabetes, damage to the immune system, disruption of hormonal function, and cancer. Seafood in the most common exposure pathway of sediment contamination to humans; therefore, for the protection of human health, the remediation of these contaminants in aquatic environments is of the upmost importance [2–5].

The first step required for remediation is an accurate assessment of the spatial distribution of contamination in order to ensure the most effective and least costly remediation. Spatially continuous data are required to delineate the boundaries of unsafe levels of contamination and to determine the volume of contaminated material to be removed. However, in aquatic environments, point samples are generally collected on a predetermined grid spacing, and the contaminant concentrations are spatially interpolated to provide a continuous surface that may introduce uncertainty. Spatial interpolation methods are traditionally grouped into deterministic and geostatistical methods. The most commonly used deterministic method is inverse distance weighting (IDW), which uses a function that estimates values at un-sampled points through the linear combination of known sample values weighted by the distance between them [6,7]. IDW is a simple method requiring minimal modeler input [8]; however, because it is based solely on distance, it often performs very poorly with sparsely distributed geospatial data. The most commonly used geostatistical method for spatial interpolation is kriging, which employs a semi-variogram that plots the semi-variance between points against the distance between them [6,9]. From the semi-variogram, it is possible to determine the range of spatial dependence of sampled points that can be used to determine a value at an unknown location [10]. However, the accurate estimation of a semi-variogram is complicated, computationally expensive, and can introduce modeler bias. Furthermore, kriging has been shown to have a significant smoothing affect, where areas of high pollutant concentration could be missed. This is not ideal as an underestimation of the pollutant concentrations could lead to an increased risk to human health. Both deterministic and geostatistical methods have one major commonality, that is, the greater the sample density, the greater the accuracy of the spatial interpolation [11–13]. However, the sampling cost of sediment in marine environments and the analytical assessment cost for dioxins are extremely high. Therefore, obtaining an adequate number of samples to achieve an acceptable resolution during interpolation may not be possible, and this high cost may be prohibitive to remediation projects.

There is a separate family of data-driven predictive methods that utilize fuzzy set theory [14], which have been proven to be a suitable method for the prediction of sparse non-linear data and have been used for many applications of spatial estimation in geoscience [11,15–17]. Fuzzy set theory [14] provides a convenient way of describing the degree of belonging (membership μ) of an element to a set, between 0 (no belonging) and 1 (complete belonging). For example, let U be an ordinary set with elements $\{x_1, x_2, \ldots, x_n\}$ and $\widetilde{A}$ be a fuzzy subset of U, in which the elements x_i have degrees of membership (belonging to $\widetilde{A}$) given by a membership function $\mu_{\widetilde{A}}(x_i) = \alpha$, which dictates that an element x_i has a degree of membership α to fuzzy set $\widetilde{A}$, where $0 \leq \alpha \leq 1$ (see Figure 1).

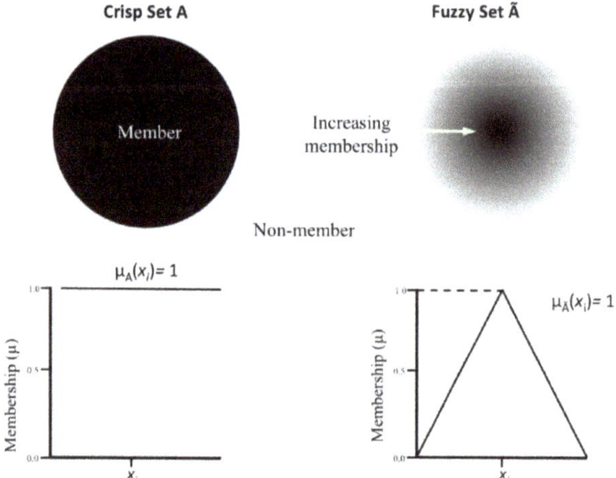

Figure 1. Example of a (**left side**) crisp and (**right side**) fuzzy set, and their respective membership functions.

The Takagi-Sugeno Method

System modeling techniques that employ fuzzy set theory are commonly referred to as fuzzy modeling [15,18,19]. One of the most commonly used fuzzy modeling techniques for spatial estimation is the Takagi–Sugeno (TS) method [15,18,20]. TS fuzzy modeling breaks down the input data space into a number of fuzzy regions and creates a linear function for each region. It is advantageous for spatial modeling because of its transparency and interpretability [21]. The degree of belonging that an un-sampled point has to each region in the data is used to predict a value at that location. The partitioning of the input space into fuzzy regions is achieved though fuzzy clustering, which is the foundation for the fuzzy spatial modeling using the TS method. Fuzzy clustering differs from traditional crisp data clustering in that in fuzzy methods each element (sample point) can have a degree of membership to multiple clusters within the data. Each cluster is defined by a cluster center, which has a value calculated from the membership-weighted average of the members of that cluster [22,23]. For spatial modeling, data are clustered in the three-dimensional product space defined by the Cartesian map coordinates (x, y) and the magnitude of a pollutant concentration (p) (see Figure 2).

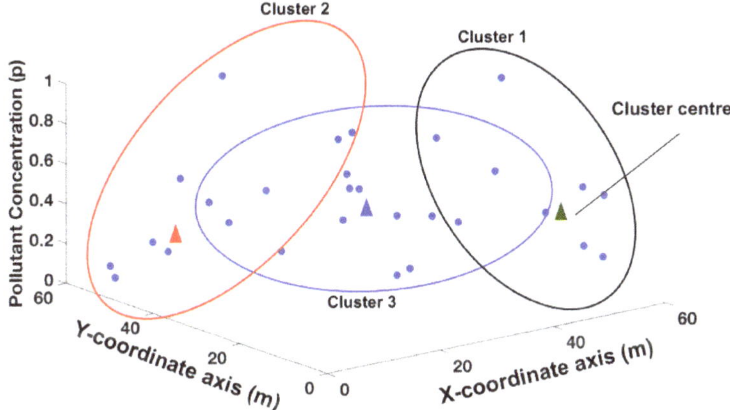

Figure 2. Theoretical visualization of fuzzy clustering in the three-dimensional product space.

After clustering, each data point has membership to one or more cluster centers, depending on the pollutant concentration. The clusters are then partitioned onto the x and y Cartesian product space and a membership function is generated for the x and y coordinate axis of each cluster. The membership functions from each cluster are then used to determine the degree of membership an un-sampled location has to the different clusters within the data (see Figure 3).

Ultimately, the combination of these degrees of membership is used in solving a pollutant concentration at an unknown location. Once the data have been clustered, they are subjected to a rule-based fuzzy inference system (FIS) that makes inferences about un-sampled geographic locations based on their membership to the clusters within the data. A single rule is introduced for each cluster using conditional "IF-THEN" statements. FIS uses input variables referred to as antecedents for each rule; in the TS fuzzy model, the antecedents are the fuzzy sets generated from the clustered data. The antecedents are subjected to IF-THEN statements to produce a weighting for the prediction from each rule based on membership to that rule (cluster). The output of each rule is referred to as a rule consequent. When a rule in the FIS is executed, if the antecedent is unaffected by the IF-THEN condition, that rule is skipped and the next rule is executed. If the IF-THEN condition produces a consequent, then that rule is deemed to have fired or been executed.

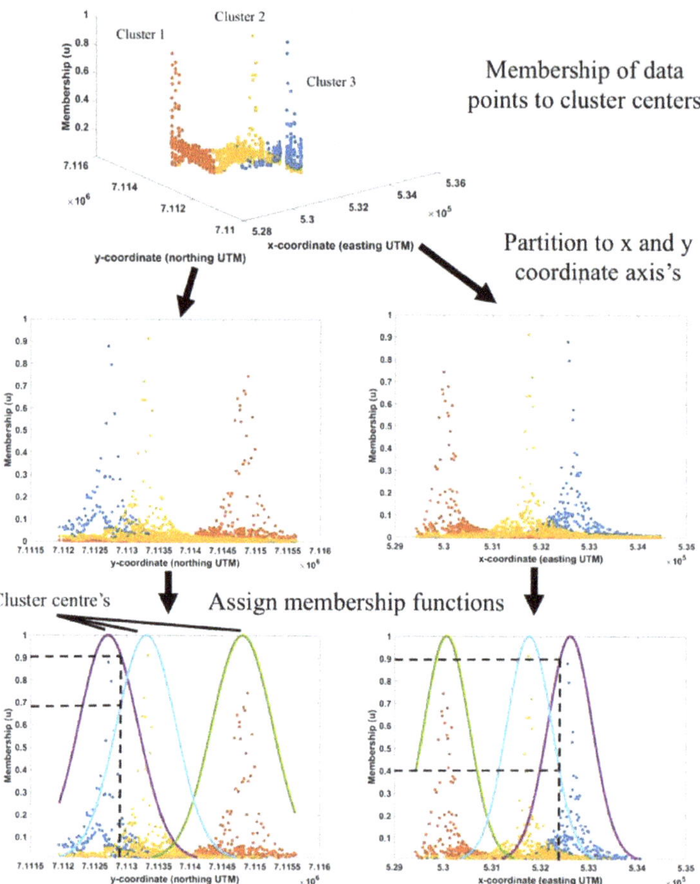

Figure 3. Example of partitioning membership to cluster centers to the map coordinate axes and applying a membership function (represented by solid lines) for three fuzzy clusters to determine membership based on geographic location.

Each rule in the TS fuzzy model uses x and y Cartesian coordinates as inputs to solve the consequent (output variable) for each rule in the form of a linear equation to produce a crisp value [23]. The form of the general first order TS model is shown in Equation (1).

$$R_i = if\ x_1 is\ A_{i1}\ and\ x_2\ is\ A_{in}\ then\ p_i = a_{i1}x_1 + a_{i2}x_2 + b_i \quad i = 1, \ldots, K \tag{1}$$

where R_i is the ith rule; x_1 and x_2 are the antecedent variables (Cartesian coordinates x, y); A_{i1} and A_{i2} are fuzzy sets for the ith rule and the respective coordinate axes; p_i is the ith rule output; K is the number of rules (clusters); and a_{i1}, a_{i2}, and b_i are the unknown model parameters that must be solved. This is accomplished by least squares regression using the data in each cluster. The model output for a given input is obtained through the aggregation of all utilized rule consequents weighted by their membership to the respective rules [15,19,20]. The TS method's simplistic nature and interpretability make it advantageous compared to other FIS [24].

Applications using TS fuzzy modeling for spatial estimation have shown its predictive capacity to outperform kriging [19,25]. The advantage of TS modeling is that it allows a complex data surface to be broken down into more easily modeled individual fuzzy surfaces. Clustering and rule-based methods are then used to estimate a value at unknown locations by solving the intersection of the contributing surfaces based on the degree of belonging an unknown location has to each surface.

Fuzzy set theory has been cited as a useful tool for modeling under sparse data conditions [26]. Fuzzy modeling can take advantage of this additional information to produce an adequate contaminant distribution map for remediation planning using fewer point samples. Lowering the number of samples required would reduce the initial cost of new remediation projects and may lead to more remediation projects being undertaken in the future, eventually leading to a cleaner environment and to lower anthropogenic health risks [27].

This research employs a spatially distributed marine, soil geochemical dataset to compare the predictive abilities of the TS fuzzy model, IDW method, and ordinary kriging method (OK) using a sparse dataset. The geochemical signatures present in the data are the result of naturally occurring mineralization and are not related to anthropogenic sources. The data are useful, because the indicator geochemical signatures associated with the mineralization are analogous with several pollutants of concern (POC) in harbour settings. The spatial distribution of the contaminates in the data are very similar to that from anthropogenic sources, as both generally originate from point sources and spread out into the surrounding environment.

2. Materials and Data

The dataset used in this research is comprised of 1535 spatially distributed sample points (illustrated in Figure 4), each consisting of a 51 element spectral array with notable POCs of arsenic, mercury, and lead. Although no two spatial datasets would have identical statistical distributions, the challenges faced during spatial interpolation are similar regardless of the data type. Therefore, the use of this data still provides an adequate initial test of the relative predicative ability of the TS fuzzy model under increasingly sparse spatial data density.

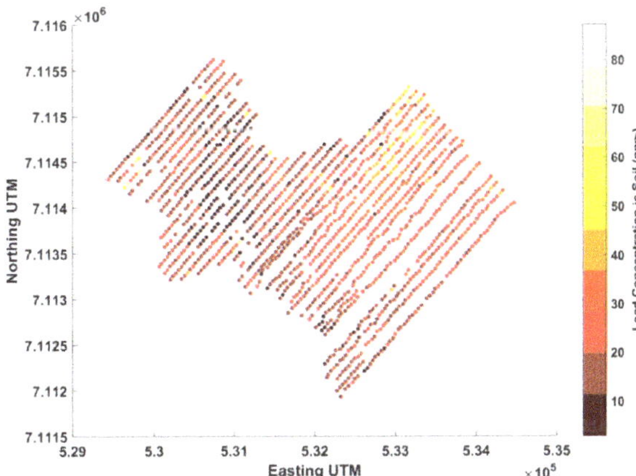

Figure 4. Spatial expression of the data used for this research, Datum: NAD83, UTM Zone 8, Eastern Yukon Territory.

The 1535 soil samples were collected on an approximately 50 m by 50 m grid spacing at depths ranging from 0.1–1.2 m below the ground surface. The samples had compositions ranging from well to poorly developed soils to glacial sediment. The sample area was approximately 5000 m × 4000 m in size and was in the Yukon Territory, Canada. The samples were collected between 17 August 2015 to 22 August 2015, and between 16 June 2016 to 25 June 2016 by a sampling team from Archer, Cathro, and Associates Limited under contract for ATAC Resources Limited. ALS Minerals in Vancouver BC analyzed the samples using metallurgical assay with inductively coupled plasma and atomic emission spectrometry. The spatial coordinates of each sample were recorded in the spatial datum NAD83 UTM Zone 8. Each data point was characterized using an easting and northing UTM coordinate in meters. Because of its relevance to human health, its high rank as a POC in urban settings and its availability in the soil data, lead was selected to test the efficacy of TS fuzzy modeling compared with OK and IDW in this work.

As OK and IDW are well-developed tools, this research employed the GIS platform ArcMap 10.4.1 (ESRI, 2011) and used the Geostatistical Analyst Package to perform the OK and IDW methods using optimized parameters. Using the spatial predictions from ArcMap, model validation was performed using MATLAB R2017a (MathWorks, 2017). The fuzzy model was also developed and validated, and its results were compared with OK and IDW using MATLAB.

In order to assess the predictive ability of the TS fuzzy, kriging, and IDW methods, the sample data were split into training and validation sets, where the training data were used to predict the known pollutant concentrations at the locations in the validation set. To assess the predictive ability of the models under increasingly sparse spatial data conditions that simulated fewer samples, the amount of training data used were incrementally reduced from 75%, to 66%, to 50%, to 33%, to 25%, to 12.5% and labelled as sets A–F, respectively. At each increment, the data not used for training were utilized for validation.

To determine the optimal number of clusters, the fuzziness parameter (m) for the FCM clustering and the membership function width (σ) for the FIS, a pseudo-optimization was performed. This was accomplished by iteratively increasing the number of clusters and at each iteration varying m through a reasonable range, and at each iteration of m varying σ through a reasonable range. The model performance results from all combinations were then assessed and the combination of m and σ that yielded the most accurate prediction for the respective validation sets were retained as optimal.

The key things that determined the competency of a spatial interpolator were its predictive ability, its ability to handle data of different types and variance, and the smoothness or abruptness of the surface generated [6]. The performance metrics used to assess the model performance of TS fuzzy, OK, and IDW modeling were the coefficient of determination (R^2), root mean squared error (RMSE), and the mean absolute error (MAE). R^2 is commonly used and provides a metric by which to assess the variance of the model prediction. A higher R^2 value indicates that a higher proportion of the total variation of the prediction is explained by the model [19]. An R^2 value equal to 1 would indicate a perfect prediction. Each model result for MAE, R^2, and RMSE received a score between 1 and 10. The mean score then determined the overall performance of that model result, with 10 being the best and 1 being the worst. The ranges for the scoring system were chosen arbitrarily based on the performance ranges observed from each metric. This approach aimed to produce enough separation between scores to draw realistic inferences on the differences in performance.

3. Results

When determining the optimal number of clusters, it was observed that the performance of the of TS fuzzy model would reach an initial peak or plateau and subsequent model runs using a higher number of clusters would not produce any further increase in score. In the interests of maintaining minimal complexity within the model and reducing the risk of over clustering, the lowest number of clusters that yielded the highest model performance was retained as optimal. In cases where more than one TS fuzzy model iteration with the optimal number of clusters but different parameters produced the same mean score, the result with the lowest mean absolute error (MAE) was deemed optimal.

The MAE was selected because of its prevalence in assessing the accuracy of spatial predictions [28]. Figure 5 displays the results from the selection of the optimal number of clusters for data reduction increments A and B for subset 1. In each case, the highest mean score from each number of clusters tested is displayed. Therefore, for each cluster, the results displayed are the optimal performance with the optimal m and σ for that number of clusters.

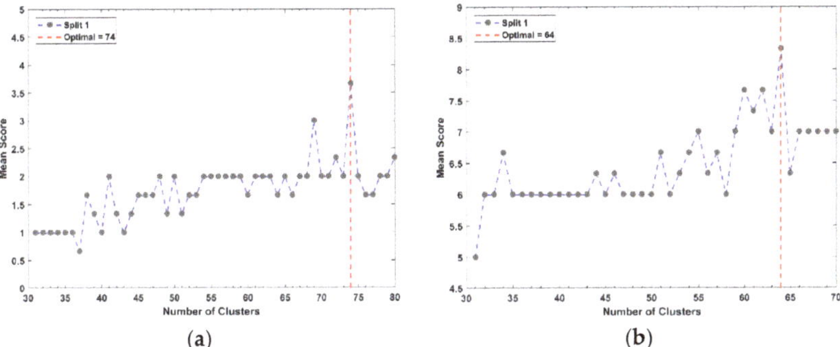

Figure 5. Example results from the determination of the optimal number of clusters for data reduction for (**a**) increment A and (**b**) increment B, for subset 1.

In all cases, a clear initial peak was reached and dictated the optimal number of clusters for that particular training set. After identifying the number of clusters that yielded the highest model performance, the combination of fuzziness (m) and membership function width (σ) that contributed to the highest mean score were identified. Figure 6 displays an example of the results from determining the optimal m value. In each case, the highest mean score for each m value tested with the optimal number of clusters is displayed.

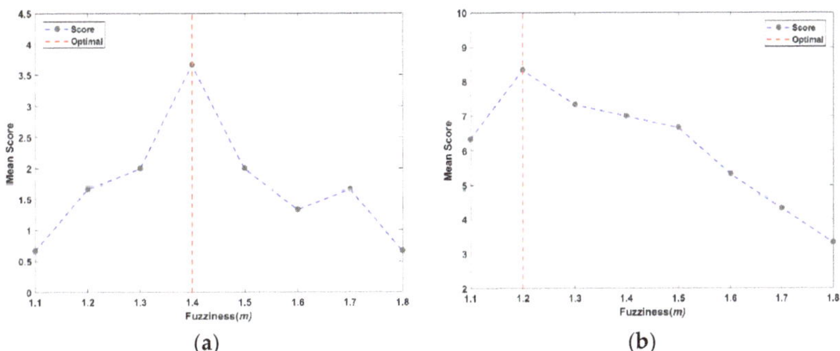

Figure 6. Example results from the determination of m using the optimal number of clusters for data reduction for (**a**) increment A and (**b**) increment B.

In all instances, a single m value produced the highest model performance for each increment of the data reduction and the m value had a visibly large effect on the performance of the TS fuzzy model. The optimal σ, which produced the highest model performance, was determined to occur over a range and was not overly sensitive. In general, a range of 80 m to 150 m produced similar mean scores. Figure 7 displays a 2D visual representation of the clustered data, the cluster centres, and the spatial distribution of membership to the cluster centres. The concentration of lead at a validation point was solved using its membership to each of the clusters in the data.

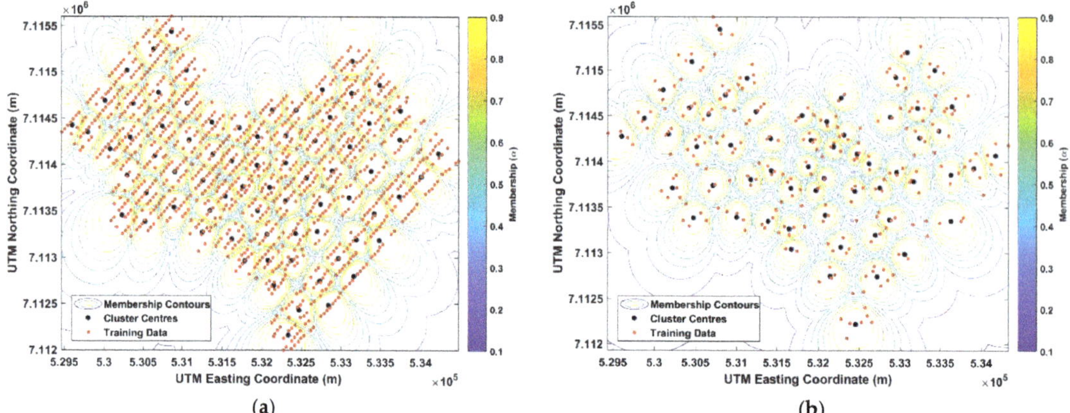

(**a**) (**b**)

Figure 7. (**a**) Visual representation of clustered data for training increment A and (**b**) training increment F.

Finally, using the optimal model parameters, the fuzzy model was trained using each increment's training data (A–F); then, the lead concentrations at the spatial locations of each increment's validation sets were solved using the model. The predictions were then compared with the actual values and the performance metrics applied. This same methodology was also applied to the OK and IDW methods, such that the performance of the three models could be compared. Figure 8 displays the performance results for each model at all of the data reduction increments.

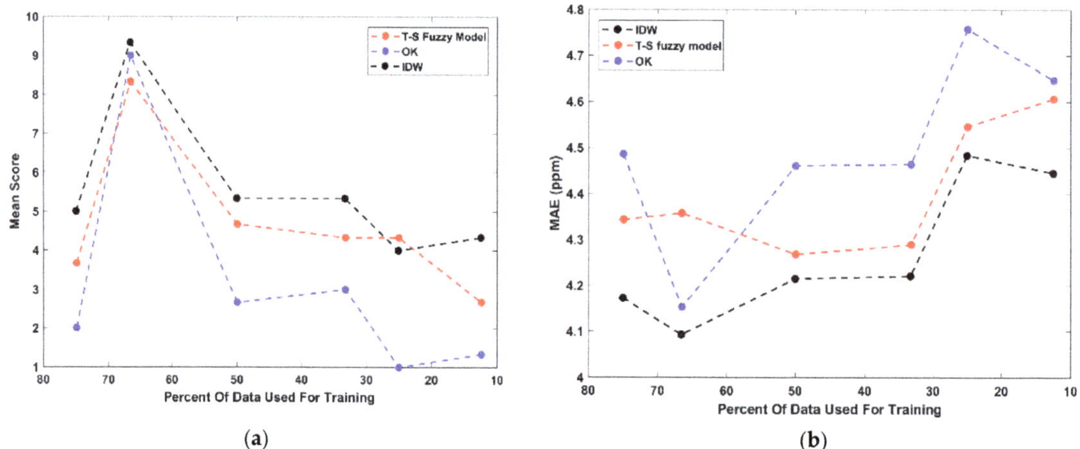

(**a**) (**b**)

Figure 8. *Cont.*

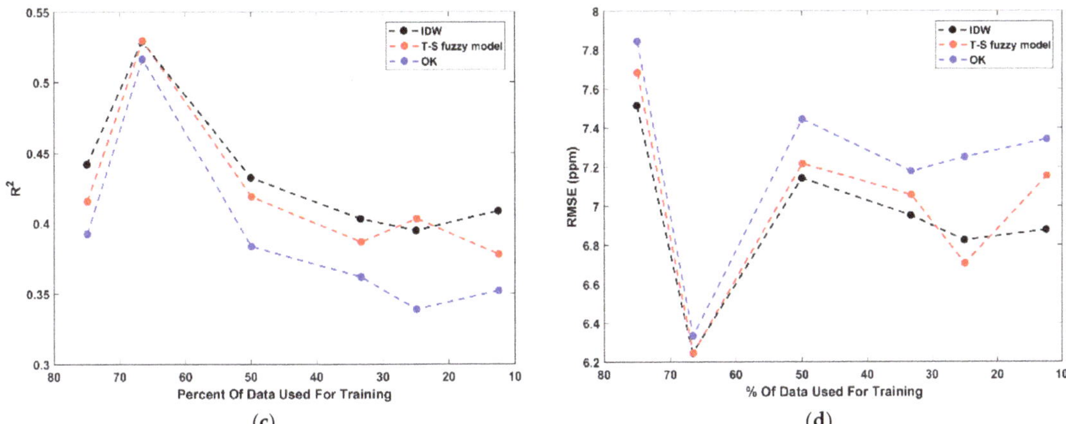

Figure 8. Ordinary kriging (OK), inverse distance weighting (IDW), and the Takagi–Sugeno (TS) fuzzy model from each increment of data reduction performance metrics (**a**) mean score, (**b**) mean absolute error (MAE), (**c**) coefficient of determination (R^2), and (**d**) root mean squared error (RMSE).

4. Discussion

In general, the three models tested here performed very similarly. For all three methods, data reduction increment B produced the highest performance scores. This is because this increment's training set has a much higher range, variance, and kurtosis than the respective validation set, making for simpler predictions closer to the mean. The discrepancies identified between the training and validation sets appear to have a large effect on the performance of the models—specifically, between increment A and B, where, for increment A, the kurtosis and range are higher in the validation set and for increment B, the kurtosis and range are higher in the training set. The effect of statistical differences between the training and validation sets does appear to be significant. However, in a real-world situation, if fewer samples are collected and analyzed, the results could have a lower range and kurtosis; so, for the purposes of simulating spatial data sparsity, the training and validation sets used here are useful because they illuminate weaknesses in the models tested. This proves that if the samples collected fail to capture the overall variance of a system, their spatial predictions will be poor, which Li and Heap (2011) found when reviewing 18 studies using different spatial interpolators.

Throughout the data reduction increments, on average, all of the methods' performances decline, which is the desired result of the evaluation. For increments A–F, IDW consistently outperformed the other methods, and the TS fuzzy model outperformed OK, but only by a small margin. To further quantify these observations, the performance of the individual metrics from the optimal TS fuzzy model results and for the OK and IDW methods from increments A–F are also plotted against the amount of training data used in each increment (shown in Figure 8). The differences between the three methods are so subtle that it is difficult to draw any significant conclusions from the results. As is consistent with the mean scores, based on these performance metrics, IDW appears to produce the best results, with the TS fuzzy model outperforming OK by a small amount. For the lead predictions made in this analysis, OK, IDW, and the TS fuzzy model all appear to have a significant smoothing effect at all data reduction increments, which agrees with previous research [12,29]. Previous research comparing IDW and OK found that in general, the two methods perform very similarly, and that IDW often produces a lower error, and this research agrees with that finding [29–31]. Furthermore, previous research has also shown that TS fuzzy modeling has the ability to outperform kriging [25,32,33] as it has in this study, albeit by a small amount. Although not clear in this research, the TS fuzzy model did produce a higher mean score than OK for data reduction increments C–F. Although the three models are not using identical distributions in their predictions of unknown points, it is possible that the similarity in the results is simply a function of the inherent similarities within the models themselves. However, further research would be required in order to quantify this.

As fuzzy clustering is at the core of the TS fuzzy modeling approach, it can be surmised that the performance of the method may be limited by the quality of the fuzzy clusters or by their spherical shape, which agrees with findings from the literature [15]. Therefore, further investigation into other clustering methods is warranted. Additionally, the data used in this research have a negatively skewed distribution. A factor that was not considered was normalizing the distribution of the data prior to the analysis, which would impact the performance of the models. However, for testing the relative performance of the TS fuzzy model against IDW and OK for spatial prediction using incrementally less spatially distributed data, the results do show that the TS fuzzy model in this research is not a superior prediction method.

A useful confirmation from this research is that fuzzy set theory is highly flexible [14] and thus, has many other applications including the prediction of time dependent variables [26]. Fuzzy-based methods lend a flexibility to environmental modelling and assessment that crisp methods do not [34,35]. It can be incorporated into other types of modelling schemes such as those that consider connectivity in earth system processes, for example. Connectivity is a subject that describes the degree to which water flow and sediment are related or connected [36,37], and can be used to acquire enough accurate data of past and current conditions to assess the role of landscape processes. When contaminants are costly to analyze, as is the case for many emerging contaminants of concern in coastal regions, Keesstra et al. [36] showed that a connectivity modelling scheme can be used to help define a more cost-effective monitoring and measurement plan.

5. Conclusions

A TS fuzzy model using the fuzzy c-means clustering algorithm was used to predict lead in soil concentrations under increasingly sparse spatial data conditions. The results from the TS fuzzy model were compared with that of OK and IDW using the same training and validation datasets. The ability of the three methods to predict outlier points within the respective validation sets appeared to be very similar. As with all other performance metrics, the performance of the three methods was difficult to separate and the similarity in model performance may be related to the models themselves. Both IDW and OK generate weights for the surrounding points based on a similar model shape, while the TS fuzzy model utilizes Gaussian shaped membership functions to determine the weighting for each of the cluster's predictions. This research showed that the TS fuzzy model used here did not outperform OK and IDW for the spatial predictions of lead in soil under increasingly sparse data conditions. Future research should consider different types of data, normalizing data prior to fuzzy modeling, other clustering methods, and further optimization of the model parameters.

Author Contributions: Conceptualization, R.T., U.T.K. and C.V.; methodology, R.T., U.T.K. and C.V.; validation, R.T.; formal analysis, R.T.; investigation, R.T.; resources, C.V.; data curation, R.T.; writing—original draft preparation, R.T. and F.T.; writing—review and editing, R.T., U.T.K., C.V. and F.T.; supervision, C.V. and U.T.K.; project administration, C.V. All authors have read and agreed to the published version of the manuscript.

Funding: This research received no external funding.

Acknowledgments: The authors would like to thank the reviewers for their insightful comments in their reviews of this work. The authors would also like to thank the Department of Mechanical Engineering at the University of Victoria for access to software used in this work.

Conflicts of Interest: The authors declare no conflict of interest.

References

1. Hites, R.A. Dioxins: An overview and history. *Environ. Sci. Technol.* **2011**, *45*, 16–20. [CrossRef]
2. Travis, C.C.; Hattemer-Frey, H.A. Human exposure to dioxin. *Sci. Total Environ.* **1991**, *104*, 97–127. [CrossRef]
3. Mitrou, P.I.; Dimitriadis, G.; Raptis, S.A. Toxic effects of 2,3,7,8-tetrachlorodibenzo-p-dioxin and related compounds. *Eur. J. Intern. Med.* **2001**, *12*, 406–411. [CrossRef]
4. Kulkarni, P.; Crespo, J.; Afonso, C. Dioxins sources and current remediation technologies—A review. *Environ. Int.* **2008**, *34*, 139–153. [CrossRef] [PubMed]
5. Cheung, W.H.; Lee, V.K.C.; McKay, G. Minimizing dioxin emissions from integrated MSW thermal treatment. *Environ. Sci. Technol.* **2007**, *41*, 2001–2007. [CrossRef]
6. Li, J.; Heap, A.D. A review of comparative studies of spatial interpolation methods in environmental sciences: Performance and impact factors. *Ecol. Inform.* **2011**, *6*, 228–241. [CrossRef]
7. Shepard, D. A two-dimensional interpolation function for irregularly-spaced data. In Proceedings of the 1968 23rd ACM National Conference, New York, NY, USA, 1968; pp. 517–524. Available online: https://dl.acm.org/doi/proceedings/10.1145/800186 (accessed on 28 May 2021). [CrossRef]
8. Shahbeik, S.; Afzal, P.; Moarefvand, P.; Qumarsy, M. Comparison between ordinary kriging (OK) and inverse distance weighted (IDW) based on estimation error. Case study: Dardevey iron ore deposit. NE Iran. *Arab. J. Geosci.* **2014**, *7*, 3693–3704. [CrossRef]
9. Matheron, G. Principles of geostatistics. *Econ. Geol.* **1963**, *58*, 1246–1266. [CrossRef]
10. Burrough, P.A.; McDonnell, R.A.; Lloyd, C.D. *Principles of Geographical Information Systems*, 3rd ed.; Oxford University Press: Oxford, NY, USA, 2015.
11. Gedeon, T.D.; Wong, K.W.; Wong, P.M.; Huang, Y. Spatial interpolation using fuzzy reasoning. *Trans. GIS* **2003**, *7*, 55–56. [CrossRef]
12. Goovaerts, P. Geostatistics in soil science: State-of-the-art and perspectives. *Geoderma* **1999**, *89*, 1–45. [CrossRef]
13. Stahl, K.; Moore, R.D.; Floyer, J.A.; Asplin, M.G.; McKendry, I.G. Comparison of approaches for spatial interpolation of daily air temperature in a large region with complex topography and highly variable station density. *Agric. For. Meteorol.* **2006**, *139*, 224–236. [CrossRef]
14. Zadeh, L.A. Fuzzy sets. *Inf. Control* **1965**, *8*, 338–353. [CrossRef]
15. Muhammad, K.; Glass, H.J. Modelling short-scale variability and uncertainty during mineral resource estimation using a novel fuzzy estimation technique. *Geostand. Geoanalytical Res.* **2011**, *35*, 369–385. [CrossRef]
16. Collazo-Cuevas, J.I.; Aceves-Fernandez, M.A.; Gorrostieta-Hurtado, E.; Pedraza-Ortega, J.C.; Sotomayor-Olmedo, A.; Delgado-Rosas, M. Comparison between fuzzy c-means clustering and fuzzy clustering subtractive in urban air pollution. In Proceedings of the 2010 20th International Conference on Electronics Communications and Computers, Cholula, Puebla, Mexico, 22–24 February 2010; pp. 174–179. [CrossRef]
17. Sonmez, H.; Gokceoglu, C.; Ulusay, R. A mamdani fuzzy inference system for the Geological Strength Index (GSI) and its use in slope stability assessments. *Int. J. Rock Mech. Min. Sci.* **2004**, *41*, 513–514. [CrossRef]
18. Kajornrit, J.; Wong, K.W.; Fung, C.C. An interpretable fuzzy monthly rainfall spatial interpolation system for the construction of aerial rainfall maps. *Soft Comput.* **2016**, *20*, 4631–4643. [CrossRef]
19. Tutmez, B.; Dag, A. Use of fuzzy logic in lignite inventory estimation. *Energy Sources Part B Econ. Plan. Policy* **2007**, *2*, 93–103. [CrossRef]
20. Kazemi, S.M.; Hosseini, S.M. Comparison of spatial interpolation methods for estimating heavy metals in sediments of Caspian Sea. *Expert Syst. Appl.* **2011**, *38*, 1632–1649. [CrossRef]
21. Pedrycz, W.; Izakian, H. Cluster-centric fuzzy modeling. *IEEE Trans. Fuzzy Syst.* **2014**, *22*, 1585–1597. [CrossRef]
22. Bezdek, J.C.; Ehrlich, R.; Full, W. FCM: The fuzzy c-means clustering algorithm. *Comput. Geosci.* **1984**, *10*, 191–203. [CrossRef]
23. Takagi, T.; Sugeno, M. Fuzzy identification of systems and its applications to modeling and control. *IEEE Trans. Syst. Man Cybern.* **1985**, *SMC-15*, 116–132. [CrossRef]
24. Zhou, S.-M.; Gan, J.Q. Low-level interpretability and high-level interpretability: A unified view of data-driven interpretable fuzzy system modelling. *Fuzzy Sets Syst.* **2008**, *159*, 3091–3131. [CrossRef]
25. Kord, M.; Moghaddam, A.A. Spatial analysis of Ardabil plain aquifer potable groundwater using fuzzy logic. *J. King Saud Univ. Sci.* **2014**, *26*, 129–140. [CrossRef]
26. Khan, U.T.; Valeo, C. A new fuzzy linear regression approach for dissolved oxygen prediction. *Hydrol. Sci. J.* **2015**, *60*, 1096–1119. [CrossRef]
27. Bárdossy, G.; Fodor, J. Traditional and new ways to handle uncertainty in geology. *Nat. Resour. Res.* **2001**, *10*, 179–187. [CrossRef]
28. Willmott, C.J.; Matsuura, K. On the use of dimensioned measures of error to evaluate the performance of spatial interpolators. *Int. J. Geogr. Inf. Sci.* **2006**, *20*, 89–102. [CrossRef]
29. Zarco-Perello, S.; Simões, N. Ordinary kriging vs inverse distance weighting: Spatial interpolation of the sessile community of Madagascar reef, Gulf of Mexico. *PeerJ* **2017**, *5*. [CrossRef] [PubMed]
30. Qiao, P.; Lei, M.; Yang, S.; Yang, J.; Guo, G.; Zhou, X. Comparing ordinary kriging and inverse distance weighting for soil as pollution in Beijing. *Environ. Sci. Pollut. Res.* **2018**, *25*, 15597–15608. [CrossRef] [PubMed]
31. Mirzaei, R.; Sakizadeh, M. Comparison of interpolation methods for the estimation of groundwater contamination in Andimeshk-Shush Plain, Southwest of Iran. *Environ. Sci. Pollut. Res.* **2016**, *23*, 2758–2769. [CrossRef]

32. Tutmez, B.; Hatipoglu, Z. Comparing two data driven interpolation methods for modeling nitrate distribution in aquifer. *Ecol. Inform.* **2010**, *5*, 311–315. [CrossRef]
33. Tutmez, B.; Tercan, A.E.; Kaymak, U. Fuzzy modeling for reserve estimation based on spatial variability. *Math. Geol.* **2007**, *39*, 87. [CrossRef]
34. Cui, X.; Yang, F.; Wang, X.; Ai, B.; Luo, Y.; Ma, D. Deep learning model for seabed sediment classification based on fuzzy ranking feature optimization. *Mar. Geol.* **2021**, *432*, 106390. [CrossRef]
35. Li, H.; Cao, Y.; Su, L. Multi-dimensional dynamic fuzzy monitoring model for the effect of water pollution treatment. *Environ. Monit. Assess.* **2019**, *191*, 352. [CrossRef] [PubMed]
36. Keesstra, S.; Nunes, J.P.; Saco, P.; Parsons, T.; Poeppl, R.; Masselink, R.; Cerdà, A. The way forward: Can connectivity be useful to design better measuring and modelling schemes for water and sediment dynamics? *Sci. Total Environ.* **2018**, *644*, 1557–1572. [CrossRef]
37. Rodrigo Comino, J.; Keesstra, S.D.; Cerdà, A. Connectivity assessment in Mediterranean vineyards using improved stock unearthing method, LiDAR and soil erosion field surveys. *Earth Surf. Process. Landf.* **2018**, *43*, 2193–2206. [CrossRef]

Article

An In Silico and In Vitro Study for Investigating Estrogenic Endocrine Effects of Emerging Persistent Pollutants Using Primary Hepatocytes from Grey Mullet (*Mugil cephalus*)

Paolo Cocci, Gilberto Mosconi and Francesco A. Palermo *

School of Biosciences and Veterinary Medicine, University of Camerino, Via Gentile III Da Varano, I-62032 Camerino, MC, Italy; paolo.cocci@unicam.it (P.C.); gilberto.mosconi@unicam.it (G.M.)
* Correspondence: francesco.palermo@unicam.it

Abstract: There is growing concern about the environmentally relevant concentrations of new emerging persistent organic pollutants, such as perfluorinated compounds and pharmaceuticals, which are found to bioaccumulate in aquatic organisms at concentrations suspected to cause reproductive toxicity due to the activation of estrogen receptor (ER) α and β subtypes. Here, we use a combined in silico and in vitro approach to evaluate the impact of perfluorononanoic acid (PFNA) and Enalapril (ENA) on grey mullet (*Mugil cephalus*) hepatic estrogen signaling pathway. ENA had weak agonist activity on ERα while PFNA showed moderate to high agonist binding to both ERs. According to these effects, hepatocytes incubation for 48 h to PFNA resulted in a concentration-dependent upregulation of ER and vitellogenin gene expression profiles, whereas only a small increase was observed in ERα mRNA levels for the highest ENA concentration. These data suggest a structure–activity relationship between hepatic ERs and these emerging pollutants.

Keywords: endocrine disruptors; *Mugil cephalus*; PFNA

Citation: Cocci, P.; Mosconi, G.; Palermo, F.A. An In Silico and In Vitro Study for Investigating Estrogenic Endocrine Effects of Emerging Persistent Pollutants Using Primary Hepatocytes from Grey Mullet (*Mugil cephalus*). *Environments* **2021**, *8*, 58. https://doi.org/10.3390/environments8060058

Academic Editor: Sílvia C. Gonçalves

Received: 14 May 2021
Accepted: 16 June 2021
Published: 18 June 2021

1. Introduction

Chemicals interfering with the endocrine system known as endocrine disrupting chemicals (EDCs) are pollutants that typically occur in aquatic environments as a result of municipal wastewater discharge, landfill leachates, and agricultural and urban runoff [1]. EDCs are considered a major cause of aquatic wildlife decline and loss of biodiversity [2]. Aquatic organisms such as fish may experience life-long exposures to EDCs and may bioaccumulate them developing a wide range of hormonal abnormalities [3]. Today, there is growing concern about the environmentally relevant concentrations of new emerging persistent pollutants, such as perfluorinated compounds (PFCs) and pharmaceuticals (e.g., contraceptives and anti-depressants), which are found to bioaccumulate in aquatic food webs at concentrations suspected to perturb neuro-endocrine processes in living organisms including humans [4,5].

PFCs are synthetic chemical compounds that due to high stabilities and low surface tensions are increasingly used in various industrial applications and common consumer products [6]. Among PFCs, the perfluoroalkylated substances (PFAS) have been discovered as global pollutants remaining most persistently in each environmental compartment [7,8]. Although PFAS are considered moderately to highly toxic, some of these (e.g., perfluorooctane sulfonic acid (PFOS) and perfluorooctanoic acid (PFOA)) are suspected endocrine disruptors and have been found to cause adverse health effects, especially reproductive toxicity, in different vertebrate models [9,10]. Similarly, pharmaceuticals that include any chemical product used by individuals or agribusiness for promoting personal and livestock health, have aroused great interest as environmental pollutants for their ecotoxicological potentials [11,12]. These compounds have been detected, unchanged or as metabolites, in wastewater, surface and drinking waters throughout the world [13–15]. Calamari et al. [16]

have defined pharmaceuticals as pseudo-persistent pollutants due to their continuous introduction into the environment, the biotic and/or abiotic transformation and the ability to exert subtle effects in non-target organisms. In this regard, the estrogenic potential of some pharmaceuticals has attracted great concern especially in the aquatic environment [17,18].

Estrogen-like EDCs (xenoestrogens) have the capability to bind to the estrogen receptors (ERs), mimicking the female steroid hormone, 17β-estradiol (E2), and thus activating intracellular signaling pathways. Activation of the ER-mediated signaling pathway has been extensively studied in several models, particularly fish in which feminization has been considered a direct result of xenoestrogen contamination [19–23]. In this regard, the ER-induced hepatic vitellogenin (Vtg) production is typically used to confirm exposure to estrogenic compounds in male fish [24,25]. Of the different fish organ cells, liver cells are widely used in in vitro primary culture models due to their ability to retain native liver properties including estrogen responsiveness [26–28]. For that reason, in vitro methods using primary cultures of fish hepatocytes represent a fundamental and recommendable alternative to in vivo studies for investigating several toxicologically relevant mechanisms [29,30].

In the present work, we focused our attention on the ability of perfluorononanoic acid (PFNA) and enalapril (ENA) to interfere with estrogen receptor signaling using a combined in silico/in vitro approach. The selected compounds belong to the most frequently detected classes of emerging EDCs in the aquatic environment such as PFAS and pharmaceuticals [31–35]. The Endocrine Disruptome program package was used to predict their potential interference with nuclear ERs in silico. A bioassay that uses primary hepatocytes from the grey mullet (*Mugil cephalus*) was then employed for screening estrogenic potential by assessing classical biomarker responses such as VTG protein and ER isoform mRNA expression. In addition, cytotoxicity using Alamar Blue and 3-(4,5-dimethylthiazol-2-yl)-2,5-diphenyltetrazolium bromide (MTT) assays was evaluated after 24 and 48 h exposure.

2. Materials and Methods

2.1. Endocrine Disruptome Screening Tool

A molecular docking approach for predicting interactions between PFNA/ENA and estrogen receptor α (ERα) and β (ERβ) ligand binding domains has been performed with Endocrine Disruptome Simulation (EDS) Tool. This web service has already been successfully adopted as a software tool for predicting the endocrine disruption potential of compounds, using well-validated crystal structures of 14 different human nuclear receptors including ER subtypes [36]. The crystal structures of 1A52, 3OLS, 1SJ0, and 1QKN have been chosen as templates on the basis of their sequence identity with fish receptors (higher than 60%). The docking scores reported are a measure of how the contaminants fit within the receptor-binding pocket, taking into account continuum and discreet parameters. According to the threshold calculations sensitivity (SE), it is possible to obtain four broad groups indicating predicted affinity for ER isoforms as follows: "red" ($SE < 0.25$), high probability; "orange" ($0.25 < SE < 0.50$) and "yellow" ($0.50 < SE < 0.75$), medium probability; and "green" ($SE > 0.75$), low probability of binding [36].

2.2. Hepatocyte Isolation and Primary Cell Culture

Flathead grey mullet (*Mugil cephalus*) males (95.5 ± 10.9 g initial weight) were provided by professional fishermen during fishing activities. Fish were acclimated for 2 weeks in 2.00 m × 2.00 m × 0.60 m tanks with constant aeration and natural photoperiod at Unità di Ricerca e Didattica of San Benedetto del Tronto (URDIS), University of Camerino in San Benedetto del Tronto (AP, Italy). Water quality parameters were monitored daily showing the following values: pH 8.4 ± 0.2, O_2 = 10.3 ± 0.5 mg L^{-1}, and temperature = 20–22 °C, salinity 36 ± 2 psu; undetectable level of nitrites and ammonia. Following the acclimation, fish were randomly euthanized using MS-222 within 5 min after capture. Animal manipulation was executed following the procedures established by the Italian law (Legislative Decree 116/1992), the European Communities Council Directive (86/609/EEC and 2010/63/EU) for animal welfare and under the supervision of the authorized investigators. The liver tissue was collected to obtain hepatocytes under a laminar flow hood, according to Cocci et al. [37] and Palermo et al. [38]. Purified hepatocytes were suspended in Leibovitz (L-15) phenol red-free medium, antibiotic-antimycotic solution (100 U/mL) and 10 mM HEPES. The cell density was measured in a Burker Chamber and the viability of hepatocytes was over 90%, as assessed with the Trypan blue exclusion assay. Cells were seeded on 24-well Falcon Primaria culture plates (1×10^6 cells per well) in L-15 phenol red-free medium, antibiotic-antimycotic solution (100 U/mL) and 10 mM HEPES. Cells were cultured for 24 h in an incubator at 23 °C before chemical exposure to allow attachment. Then, 50% of the L-15 phenol red-free medium culture was removed, and hepatocytes were exposed to medium containing the vehicle (ethanol, final concentration 0.01%) and 1.0, 0.01, or 0.0001 µM of E2, ENA or PFNA. Hepatocytes were incubated in an incubator at 23 °C for 96 h. Media and cells were harvested separately at 0, 24, 48, 72 and 96 h with medium changes every 24 h. Doses of ENA and PFNA were chosen on the basis of environmentally relevant concentrations [33,39–41] and six independent wells were setup for both the control and each concentration of compound. The entire experiment was repeated 3 times.

2.3. *MTT Cytotoxicity Assay*

The 3-(4,5-dimethylthiazol-2-yl)-2,5-diphenyl-tetrazoliumbromide (MTT) activity was measured according to Smeets et al. [42] with slight modifications, using the MTT Cell Proliferation and Cytotoxicity Assay Kit assay (Boster Biological Technology, Pleasanton, CA, USA, Catalog # AR1156). After 0, 24, 48, 72, 96 h of treatment described above, incubation medium was removed and replaced with fresh culture medium containing MTT reagent (5 mg/mL MTT diluted in phosphate buffered saline, PBS). After an incubation of 40 min at 23 °C, the formazan crystals produced were solubilized by adding 200 µL Formazan solubilization solution. After the complete solubilization, 200 µL of medium was transferred to a 96-well microplate and absorbance values were measured at 570 nm using a microplate reader (BioChrom, Cambridge, UK).

2.4. *Alamar Blue Assay*

Cell viability was also quantified using the Alamar Blue™ assay reagent (Thermo Scientific, Waltham, MA, USA) as described by Cocci et al. [23] and following manufacturer's specifications. The incubation medium was removed after 24, 48, 72, 96 h of treatment, replaced with a fresh culture medium containing AB reagent at a final concentration of 10%, and incubated for an additional 1 h. The absorbance was monitored at 570/600 nm using a microplate reader. The cell viability was normalized to that of hepatocytes cultured in the regular media without any of the tested compounds.

2.5. *Quantitative Realtime PCR (q-PCR)*

After exposure, the medium was carefully removed, and cells were lysed by adding the TRIzol reagent (Invitrogen Life Technologies, Milan, Italy). Total RNA was isolated according to the manufacturer's specifications. RNA quality and concentration were measured spectrophotometrically at 260/280 nm, and purity was confirmed by electrophoresis through 1% agarose gels stained with SafeView Classic (abm). The cDNA was synthesized from 1.5 µg of total RNA in 20 µL using the 5X All-In-One RT MasterMix (with AccuRT Genomic DNA Removal Kit) according to manufacturer's instructions (abm). SYBR green-based real-time PCR was used to evaluate expression profiles of ERα, ERβ, VTG target genes. 18s rRNA was selected as appropriate reference gene [28,43,44]. All the primer sequences are reported in Table 1 and were provided from Ribecco et al. [45], Vieira et al. [46], Cabas et al. [47], and Perez-Sanchez et al. [48]. The reaction included 10 µL of 2X BlasTaq™ qPCR MasterMix (abm), 0.5 µL each of forward and reverse primers (10 µM), 2 µL of cDNA template, and nuclease-free H_2O to a final volume of 20 µL. The expression of individual gene targets was analyzed using the ABI 7300 Real-Time PCR software. Thermo-cycling for all reactions was for 3 min at 95 °C, followed by 40 cycle of 15 s at 95 °C, and 60 s at 60 °C. Dissociation curve analysis revealed that a single peak was generated during the reaction demonstrating the production of a single product. Each amplified fragment was then compared with that obtained from amplification of *Sparus aurata* cDNA and verified with agarose gel electrophoresis (for details see Figure S1 in Supplementary Material). The efficiency of qPCR primer sets was reported in Table 1. Results were calculated using the $2^{-\Delta\Delta Ct}$ method and reported as fold change corrected for 18s rRNA and with respect to vehicle levels. Values are the mean ± SD of three independent experiments.

Table 1. List of primers used in this study.

Gene	Primer Sequence (5′–3′)	Genbank	Product Size	Efficiency (%)
ERα	CTGGTGCCTTCTCTTTTTGC TGTCTGATGTGGGAGAGCAG	AF136979	181	96.85
ERβ	TGTCATCGGGCGGGAAGG GCTCTTACGGCGGTTCTTGTCT	AF136980	188	91.74
VTG	CTGCTGAAGAGGGACCAGAC TTGCCTGCAGGATGATGATA	AF210428	158	96.31
18s rRNA	GCATTTATCAGACCCAAAACC AGTTGATAGGGCAGACATTCG	AY993930	135	98.65

2.6. Enzyme-Linked Immunosorbent Assay (ELISA)

VTG concentrations in the culture medium of *Mugil cephalus* hepatocytes were determined using an ELISA method previously published [25]. Cell culture media were diluted 1:8 as reported for routinely diluted media samples by Navas and Segner [49]. All samples were analyzed in triplicate. Absorbance was recorded at 492 nm using a microplate reader (Biochrom).

2.7. Statistical Analysis

Data were assessed with Graphpad prism v6.01 software (GraphPad Software Inc., San Diego, CA, USA) and expressed as mean ± standard error of the mean (SEM). Statistical analysis was performed using ANOVA (one-way analysis of variance) followed by Bonferroni's multiple comparison test. Differences with $p < 0.05$ were considered statistically significant.

3. Results and Discussion

The prediction results obtained with the EDS model for ENA and PFNA are given in Table 2.

ENA is a drug of the class of angiotensin-converting enzyme inhibitors (ACEI) that is mainly used in the treatment of arterial hypertension. Several studies indicate a beneficial interaction between ACEI and estrogens which in turn are involved in reducing ACE mRNA concentrations [50]. Zilberman et al. [51] showed that chronic exposure to ENA significantly up-regulated ERα and β protein expression in rats. To date, however, there is no experimental evidence that ACEI can bind directly to ERs. Thus, it is not surprising that, according to EDS simulation, ENA presents moderate binding affinity against the agonist-active conformation of the ERα and low affinity for both conformations of the ERβ, respectively.

PFNA is one of the three main long chain PFCs, primarily used as an emulsifier for producing fluoropolymers, that can be found at high concentrations in the environment [52]. PFNA has been detected in various waters and animal tissues worldwide suggesting high bioaccumulation potential in the food chain [53,54]. In mammalian studies, an interference with gonadal development in neonatal mice was observed after gestational exposure to PFNA [55]. In addition, an increase in estrogenic activity was reported for exposure to different PFCs, including PFNA, in in vivo studies using fish as models [56,57]. The results of in silico analysis obtained for PFNA allow us to support these effects because the simulation tool predicts its agonist activity on the ERs, showing a high probability of binding on ERβ and a slightly below probability on ERα. Only the agonist activity on the ERα from different species was previously described in the literature [56]. This latter paper reported that PFNA docked into the LBD region of ERα working as in vitro weak binders and activators. In the present study, PFNA docking scores for ERβ were more favorable than those for the ERα. In addition, an antagonist activity on the ERβ was also predicted but any evidence of this potential activity has been found in the literature.

Table 2. Prediction affinities for 17β-Estradiol (E2), Enalapril (ENA) and perfluorononanoic acid (PFNA). +/− indicate the crystal structures of estrogen receptors (ER) isoforms in complex with their respective agonist (+) or antagonist ligands (−).

CAS	Name	Structure	Receptor/Predictions Free Binding Energies (kcal mol^{-1})
50-28-2	E2		ERα+ (−10.2) ERα− (−10.0) ERβ+ (−9.6) ERβ− (−8.8)
75847-73-3	ENA		ERα+ (−8.3) ERα− (−7.7) ERβ+ (−6.8) ERβ− (−7.8)
375-95-1	PFNA		ERα+ (−8.9) ERα− (−9.1) ERβ+ (−9.7) ERβ− (−8.9)

In order to determine whether a structure–activity relationship between ERs and tested compounds was evident, primary cultures of grey mullet hepatocytes were used to investigate the impact of PFNA and ENA exposure on ERα/β and VTG expression. In the first stage of our study, we examined whether the tested chemical compounds affected cellular viability by using common in vitro cytotoxicity assays such as Alamar Blue and MTT. The cell viability was expressed as metabolic activity, displaying 100% viability in the media of both negative (EtOH) and positive control (E2) at each time point (Figure 1A).

In contrast, a significant inhibition of metabolic activity was obtained for both ENA and PFNA. We found that cell viability decreased in hepatocytes after exposure to the highest doses of ENA for 72 h (84.8% and 81.1% of solvent control at 0.01 and 1 μM ENA, respectively) and 96 h (89.3% and 65.0% at 0.01 and 1 μM ENA, respectively) (Figure 1B). Similarly, exposure of hepatocytes to 0.01–1 μM PFNA induced a significant change in viability at both 72 h (cell viability 86.7% and 67.5%, respectively) and 96 h (cell viability 71.7% and 53.9%, respectively) after the start of treatment (Figure 1C). Complete cell death was confirmed by microscopic examination of cells exposed to the tested chemicals. The observed effects on cell viability were further investigated using the MTT assay (Figure 2). ENA and PFNA performed similarly in both assays, showing the most significant decrease in hepatocyte viability (from 71.7% to 53.9%) exposed to the highest concentrations (0.01 and 1 μM) for 96 h (Figure 2B,C). However, the MTT assay failed to detect cytotoxicity for PFNA at 0.01 μM after 72 h exposure, thus proving to be slightly less sensitive than the Alamar Blue assay.

To our knowledge, only one study has shown that ENA elicits specific cytotoxic effects in primary cultures of hepatocytes through the involvement of a glutathione-dependent detoxification pathway [58]. This effect was observed at concentrations ranging from 0.5 to 2 mM in an in vitro rat cell model. The potential cytotoxic and antiproliferative effects of ENA were also found at concentration- and time-dependent manners in human HL60 acute promyelocytic leukaemia cells [59]. Interestingly, viability of ENA-treated HL60 cells was observed to drop by about 20% after exposure to 3 μM for 48 h. On the other hand, the effect of PFNA on hepatocyte viability was previously demonstrated in humans, using the WST-1 assay.

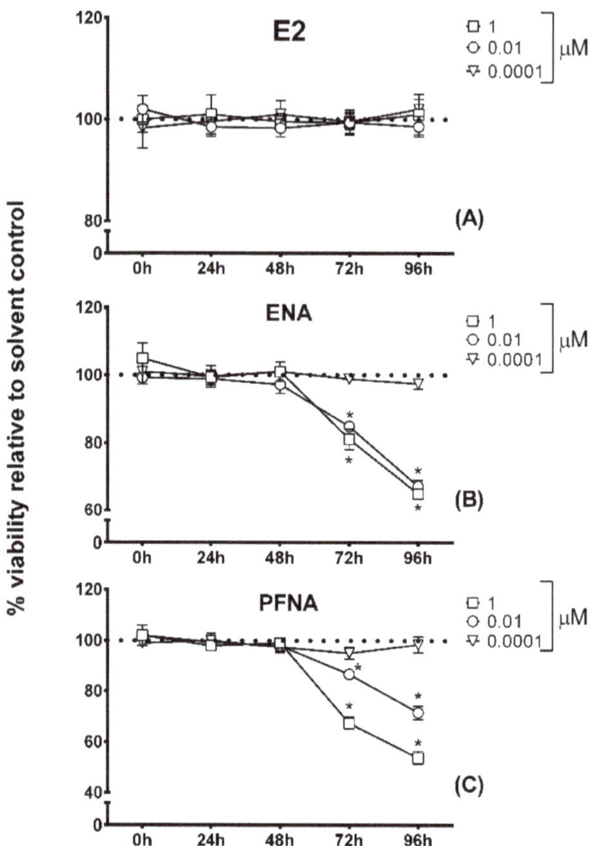

Figure 1. Alamar Blue cell viability of *Mugil cephalus* hepatocytes following exposure to E2 (**A**), ENA (**B**), PFNA (**C**) for up to 96 h. The dot line represents cell viability measured in the solvent control (assigned a survival of 100%). Values are given as mean ± SEM of three independent experiments and expressed as % relative to the solvent control. "*" indicates significant differences between control and treated groups ($p < 0.05$).

PFNA was found to be more cytotoxic than PFOA and PFOS, causing a larger decrease in cell viability upon exposure for 6–72 h to a concentration range of 200– 400 μM [60]. According to the obtained results, treatment incubation for 24 or 48 h was applied in the further studies. However, the exposure duration and sampling time of 24 h for all molecular endpoints were found to be inappropriate to obtain a clear concentration response of the model compounds (Figure 3A–C). As expected, 48 h exposure to positive control (E2) produced a dose-dependent induced expression of ERs and VTG compared with control hepatocytes (Figure 3D). In contrast, a partial concentration-response curve (0.01–1 μM) for expression of all molecular endpoints was obtained for the model compound PFNA after 48 h of exposure (Figure 3E), whereas only a small increase was observed in ERα mRNA levels for 1 μM ENA (Figure 3F).

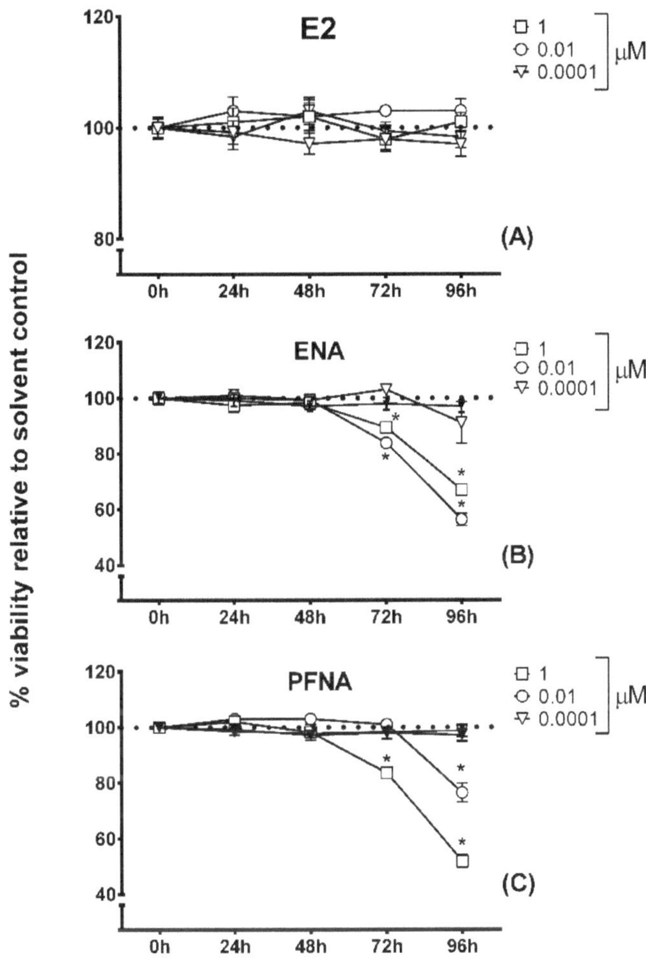

Figure 2. MTT assay of *Mugil cephalus* hepatocytes following exposure to E2 (**A**), ENA (**B**), PFNA (**C**) for up to 96 h. The dot line represents cell viability measured in the solvent control (assigned a survival of 100%). Values are given as mean ± SEM of three independent experiments in % relative to the solvent control. "*" indicates significant differences between control and treated groups ($p < 0.05$).

We further discuss the mRNA expression of VTG at the level of protein concentrations in the medium used for the primary cultures of grey mullet hepatocytes. At 48 h after treatment, E2 caused a significant dose-dependent increase in VTG synthesis at any of the tested concentrations relative to control cultures (Figure 4). Both PFNA and ENA also increased the VTG synthesis, but to a lesser extent. Indeed, a significant increase in VTG levels occurred following treatment with the highest doses (0.01 and 1 μM) of PFNA. In contrast, VTG up-regulation was only induced in response to exposure at 1 μM ENA.

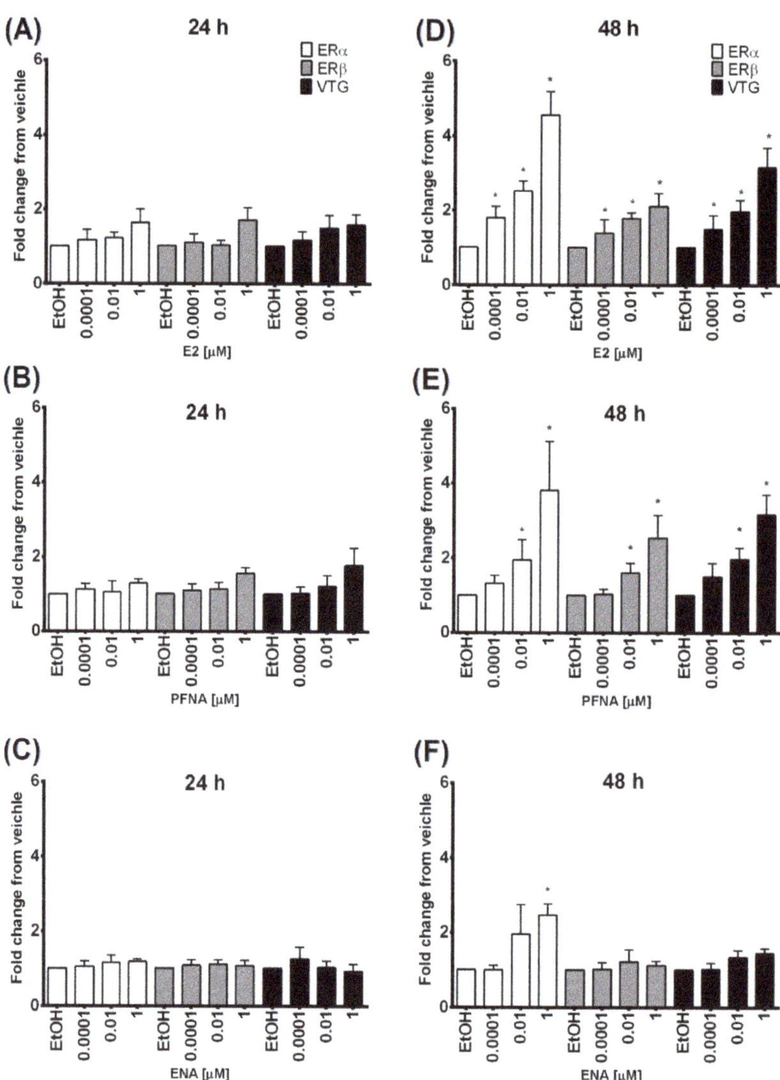

Figure 3. ERα, ERβ and VTG mRNA expression profile (fold change from vehicle) in *Mugil cephalus* hepatocytes exposed to different concentrations (0.0001–1 μM) of E2 (**A**,**D**), PFNA (**B**,**E**), ENA (**C**,**F**) for 24 and 48 h. Values are mean ± SEM of three independent experiments. "*" indicates significant differences between control and treated groups ($p < 0.05$).

ER-mediated production of VTG is likewise by far the most used biomarker of xenoestrogen exposure in oviparous species [61–65]. VTG protein and gene expression has been shown to be up-regulated by various environmental pollutants in a number of in vitro–in vivo studies using fish models [66–68]. Interestingly, most of these works have found that VTG induction is accompanied by a clear increase in hepatic ER expression, mainly ERα. This latter is indeed considered the ER subtype with a major role in mediating VTG gene induction. However, there is evidence that ERβ subtype may have a functional role in the up-regulation of ERα enhancing hepatic VTG induction in response to E2 or xenoestrogen stimulation [69,70]. Our results on PFNA are in agreement with those of Benninghoff et al. [56] who, in a recent study, found both in vitro and in vivo weak estrogenic activities of this PFC using a similar range of concentrations. Indeed, according to the relative binding affinity (RBA) values obtained in vitro, dietary PFNA also induced a consistent in vivo VTG induction in trout [56]. Collectively, our findings provide clear confirmation of data reported so far regarding the ability of PFNA to act as weak environmental xenoestrogen. PFNA has frequently been detected in surface waters at concentrations in the order of ng/L showing particularly high levels (up to a max of 100 ng/L) in the recycling sites due to the recycling activities of electrical and electronic waste [40,71]. Thus, the prominence of PFNA in water from this typology of sites suggests the need of a careful monitoring of this potential xenoestrogen in order to reduce its ecological impact in aquatic ecosystems.

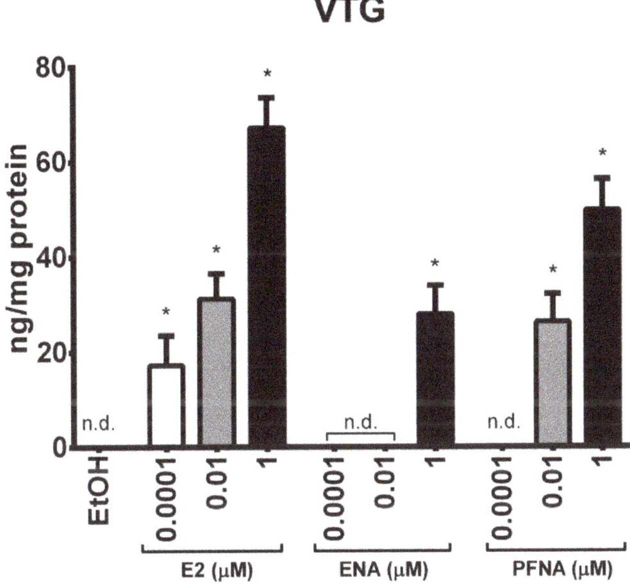

Figure 4. Changes in medium VTG levels in *Mugil cephalus* hepatocytes exposed to different concentrations (0.0001–1 μM) of E2, ENA, PFNA for 48 h. n.d.: not detectable. "*" indicates significant differences between control and treated groups ($p < 0.05$).

Similarly, the occurrence of ENA in the environment can be related to incomplete removal of this drug from wastewater treatment plants (WWTPs). ENA was detected in wastewater influent at concentrations ranging from 35 to 1400 ng/L, while in the wastewater effluent this range was reduced to 0.85–290 ng/L [39]. These data demonstrate that total or high removal of this drug can be achieved in all WWTPs, thus suggesting a substantially lower accumulation rate. Given also our results about the weak estrogenic potential, one might predict that ENA has a mild impact on reproductive functions of aquatic vertebrates.

4. Conclusions

In summary, an in vitro hepatocyte bioassay was used to characterize estrogenic responses of gray mullet to PFNA and ENA as representative compounds of two classes of emerging pollutants. According to the environmental occurrence of these chemicals, the results would indicate potential adverse impacts especially on reproductive health. The observed effects are likely to be mediated through direct actions of these compounds on hepatic ERs suggesting a structure–activity relationship between ER and these emerging pollutants. While the extent of PFNA estrogenic potential is substantially supported by literature data, further studies such as in vivo investigations need to obtain more information on the estrogenic activity of ENA, especially to check its effectiveness following long-term accumulation.

Supplementary Materials: The following are available online at https://www.mdpi.com/article/10.3390/environments8060058/s1, Figure S1. Results of PCR amplification using template DNA from *Mugil cephalus* or *Sparus aurata*.

Author Contributions: Conceptualization, P.C. and F.A.P.; methodology, P.C., G.M., F.A.P.; formal analysis, F.A.P.; investigation, P.C., G.M.; resources, F.A.P.; writing—original draft preparation, P.C., F.A.P.; writing—review and editing, F.A.P.; supervision, F.A.P. All authors have read and agreed to the published version of the manuscript.

Funding: This research received no external funding.

Informed Consent Statement: Not applicable.

Data Availability Statement: The data presented in this study are available in the article/supplementa material, further inquiries can be directed to the corresponding author.

Conflicts of Interest: The authors declare no conflict of interest.

References

1. Ng, A.; Weerakoon, D.; Lim, E.; Padhye, L.P. Fate of environmental pollutants. *Water Environ. Res.* **2019**, *91*, 1294–1325. [CrossRef] [PubMed]
2. Zhou, J.; Cai, Z.-H.; Zhu, X.-S. Are endocrine disruptors among the causes of the deterioration of aquatic biodiversity? *Integr. Environ. Assess. Manag.* **2010**, *6*, 492–498. [CrossRef] [PubMed]
3. Celino-Brady, F.T.; Lerner, D.T.; Seale, A.P. Experimental Approaches for Characterizing the Endocrine-Disrupting Effects of Environmental Chemicals in Fish. *Front. Endocrinol.* **2021**, *11*, 619361. [CrossRef] [PubMed]
4. Corsini, E.; Luebke, R.W.; Germolec, D.R.; DeWitt, J.C. Perfluorinated compounds: Emerging POPs with potential immunotoxicity. *Toxicol. Lett.* **2014**, *230*, 263–270. [CrossRef]
5. Wilkinson, J.L.; Hooda, P.S.; Barker, J.; Barton, S.; Swinden, J. Ecotoxic pharmaceuticals, personal care products, and other emerging contaminants: A review of environmental, receptor-mediated, developmental, and epigenetic toxicity with discussion of proposed toxicity to humans. *Crit. Rev. Environ. Sci. Technol.* **2015**, *46*, 336–381. [CrossRef]
6. OECD. *Summary Report on the New Comprehensive Global Database of Per- and Polyfluoroalkyl Substances (PFASs)*; Publications Series on Risk Management: Paris, France, 2018; No. 39.
7. Houde, M.; Martin, J.W.; Letcher, R.J.; Solomon, K.R.; Muir, D.C.G. Biological Monitoring of Polyfluoroalkyl Substances: A Review. *Environ. Sci. Technol.* **2006**, *40*, 3463–3473. [CrossRef]
8. Valsecchi, S.; Rusconi, M.; Polesello, S. Determination of perfluorinated compounds in aquatic organisms: A review. *Anal. Bioanal. Chem.* **2012**, *405*, 143–157. [CrossRef]
9. White, S.S.; Fenton, S.E.; Hines, E.P. Endocrine disrupting properties of perfluorooctanoic acid. *J. Steroid Biochem. Mol. Biol.* **2011**, *127*, 16–26. [CrossRef]
10. Zeng, Z.; Song, B.; Xiao, R.; Zeng, G.; Gong, J.; Chen, M.; Xu, P.; Zhang, P.; Shen, M.; Yi, H. Assessing the human health risks of perfluorooctane sulfonate by in vivo and in vitro studies. *Environ. Int.* **2019**, *126*, 598–610. [CrossRef]
11. Fent, K.; Weston, A.A.; Caminada, D. Ecotoxicology of human pharmaceuticals. *Aquat. Toxicol.* **2006**, *76*, 122–159. [CrossRef]
12. Liu, N.; Jin, X.; Feng, C.; Wang, Z.; Wu, F.; Johnson, A.C.; Xiao, H.; Hollert, H.; Giesy, J.P. Ecological risk assessment of fifty pharmaceuticals and personal care products (PPCPs) in Chinese surface waters: A proposed multiple-level system. *Environ. Int.* **2020**, *136*, 105454. [CrossRef] [PubMed]
13. Pereira, A.; Silva, L.; Laranjeiro, C.; Lino, C.; Pena, A. Selected Pharmaceuticals in Different Aquatic Compartments: Part I—Source, Fate and Occurrence. *Molecules* **2020**, *25*, 1026. [CrossRef]
14. Patel, M.; Kumar, R.; Kishor, K.; Mlsna, T.; Pittman, C.U., Jr.; Mohan, D. Pharmaceuticals of Emerging Concern in Aquatic Systems: Chemistry, Occurrence, Effects, and Removal Methods. *Chem. Rev.* **2019**, *119*, 3510–3673. [CrossRef]
15. Zuccato, E.; Castiglioni, S.; Fanelli, R.; Reitano, G.; Bagnati, R.; Chiabrando, C.; Pomati, F.; Rossetti, C.; Calamari, D. Pharmaceuticals in the Environment in Italy: Causes, Occurrence, Effects and Control. *Environ. Sci. Pollut. Res.* **2006**, *13*, 15–21. [CrossRef]
16. Calamari, D.; Zuccato, E.; Castiglioni, S.; Bagnati, R.; Fanelli, R. Strategic Survey of Therapeutic Drugs in the Rivers Po and Lambro in Northern Italy. *Environ. Sci. Technol.* **2003**, *37*, 1241–1248. [CrossRef]
17. Isidori, M.; Bellotta, M.; Cangiano, M.; Parrella, A. Estrogenic activity of pharmaceuticals in the aquatic environment. *Environ. Int.* **2009**, *35*, 826–829. [CrossRef] [PubMed]
18. Laurenson, J.P.; Bloom, R.A.; Page, S.; Sadrieh, N. Ethinyl Estradiol and Other Human Pharmaceutical Estrogens in the Aquatic Environment: A Review of Recent Risk Assessment Data. *AAPS J.* **2014**, *16*, 299–310. [CrossRef]

19. Cocci, P.; Mozzicafreddo, M.; Angeletti, M.; Mosconi, G.; Palermo, F.A. In silico prediction and in vivo analysis of antiestrogenic potential of 2-isopropylthioxanthone (2-ITX) in juvenile goldfish (Carassius auratus). *Ecotoxicol. Environ. Saf.* **2016**, *133*, 202–210. [CrossRef]
20. Cocci, P.; Palermo, F.A.; Quassinti, L.; Bramucci, M.; Miano, A.; Mosconi, G. Determination of estrogenic activity in the river Chienti (Marche Region, Italy) by using in vivo and in vitro bioassays. *J. Environ. Sci.* **2016**, *43*, 48–53. [CrossRef]
21. Palermo, F.A.; Cocci, P.; Angeletti, M.; Polzonetti-Magni, A.; Mosconi, G. PCR–ELISA detection of estrogen receptor β mRNA expression and plasma vitellogenin induction in juvenile sole (Solea solea) exposed to waterborne 4-nonylphenol. *Chemosphere* **2012**, *86*, 919–925. [CrossRef] [PubMed]
22. Palermo, F.A.; Mosconi, G.; Angeletti, M.; Polzonetti-Magni, A.M. Assessment of Water Pollution in the Tronto River (Italy) by Applying Useful Biomarkers in the Fish Model Carassius auratus. *Arch. Environ. Contam. Toxicol.* **2008**, *55*, 295–304. [CrossRef] [PubMed]
23. Cocci, P.; Capriotti, M.; Mosconi, G.; Palermo, F.A. Effects of endocrine disrupting chemicals on estrogen receptor alpha and heat shock protein 60 gene expression in primary cultures of loggerhead sea turtle (*Caretta caretta*) erythrocytes. *Environ. Res.* **2017**, *158*, 616–624. [CrossRef] [PubMed]
24. Blazer, V.S.; Walsh, H.L.; Shaw, C.H.; Iwanowicz, L.R.; Braham, R.P.; Mazik, P.M. Indicators of exposure to estrogenic compounds at Great Lakes Areas of Concern: Species and site comparisons. *Environ. Monit. Assess.* **2018**, *190*, 1–19. [CrossRef]
25. Cocci, P.; Palermo, F.A.; Pucciarelli, S.; Miano, A.; Cuccioloni, M.; Angeletti, M.; Roncarati, A.; Mosconi, G. Identification, partial characterization, and use of grey mullet (*Mugil cephalus*) vitellogenins for the development of ELISA and biosensor immunoassays. *Int. Aquat. Res.* **2019**, *11*, 389–399. [CrossRef]
26. Tollefsen, K.-E.; Bratsberg, E.; Bøyum, O.; Finne, E.F.; Gregersen, I.K.; Hegseth, M.; Sandberg, C.; Hylland, K. Use of fish in vitro hepatocyte assays to detect multi-endpoint toxicity in Slovenian river sediments. *Mar. Environ. Res.* **2006**, *62*, S356–S359. [CrossRef]
27. Tollefsen, K.-E.; Mathisen, R.; Stenersen, J. Induction of vitellogenin synthesis in an Atlantic salmon (*Salmo salar*) hepatocyte culture: A sensitivein vitrobioassay for the oestrogenic and anti-oestrogenic activity of chemicals. *Biomarkers* **2003**, *8*, 394–407. [CrossRef]
28. Cocci, P.; Capriotti, M.; Mosconi, G.; Campanelli, A.; Frapiccini, E.; Marini, M.; Caprioli, G.; Sagratini, G.; Aretusi, G.; Palermo, F.A. Alterations of gene expression indicating effects on estrogen signaling and lipid homeostasis in seabream hepatocytes exposed to extracts of seawater sampled from a coastal area of the central Adriatic Sea (Italy). *Mar. Environ. Res.* **2017**, *123*, 25–37. [CrossRef]
29. Pesonen, M.; Andersson, T.B. Fish primary hepatocyte culture; an important model for xenobiotic metabolism and toxicity studies. *Aquat. Toxicol.* **1997**, *37*, 253–267. [CrossRef]
30. Ellesat, K.S.; Yazdani, M.; Holth, T.F.; Hylland, K. Species-dependent sensitivity to contaminants: An approach using primary hepatocyte cultures with three marine fish species. *Mar. Environ. Res.* **2011**, *72*, 216–224. [CrossRef]
31. Zuccato, E.; Castiglioni, S.; Fanelli, R. Identification of the pharmaceuticals for human use contaminating the Italian aquatic environment. *J. Hazard. Mater.* **2005**, *122*, 205–209. [CrossRef]
32. Burcea, A.; Boeraş, I.; Mihuţ, C.-M.; Bănăduc, D.; Matei, C.; Curtean-Bănăduc, A. Adding the Mureş River Basin (Transylvania, Romania) to the List of Hotspots with High Contamination with Pharmaceuticals. *Sustainability* **2020**, *12*, 10197. [CrossRef]
33. Fick, J.; Söderström, H.; Lindberg, R.H.; Phan, C.; Tysklind, M.; Larsson, D.J. Contamination of Surface, Ground and Drinking Water from Pharmaceutical Production. *Environ. Toxicol. Chem.* **2009**, *28*, 2522–2527. [CrossRef]
34. Fliedner, A.; Rüdel, H.; Dreyer, A.; Pirntke, U.; Koschorreck, J. Chemicals of emerging concern in marine specimens of the German Environmental Specimen Bank. *Environ. Sci. Eur.* **2020**, *32*, 1–17. [CrossRef]
35. Zhang, X.; Lohmann, R.; Sunderland, E.M. Poly- and Perfluoroalkyl Substances in Seawater and Plankton from the Northwestern Atlantic Margin. *Environ. Sci. Technol.* **2019**, *53*, 12348–12356. [CrossRef] [PubMed]
36. Kolšek, K.; Mavri, J.; Dolenc, M.S.; Gobec, S.; Turk, S. Endocrine Disruptome—An Open Source Prediction Tool for Assessing Endocrine Disruption Potential through Nuclear Receptor Binding. *J. Chem. Inf. Model.* **2014**, *54*, 1254–1267. [CrossRef] [PubMed]
37. Cocci, P.; Mosconi, G.; Arukwe, A.; Mozzicafreddo, M.; Angeletti, M.; Aretusi, G.; Palermo, F.A. Effects of Diisodecyl Phthalate on PPAR: RXR-Dependent Gene Expression Pathways in Sea Bream Hepatocytes. *Chem. Res. Toxicol.* **2015**, *28*, 935–947. [CrossRef] [PubMed]
38. Palermo, F.A.; Cocci, P.; Mozzicafreddo, M.; Arukwe, A.; Angeletti, M.; Aretusi, G.; Mosconi, G. Tri-m-cresyl phosphate and PPAR/LXR interactions in seabream hepatocytes: Revealed by computational modeling (docking) and transcriptional regulation of signaling pathways. *Toxicol. Res.* **2015**, *5*, 471–481. [CrossRef]
39. Celiz, M.D.; Pérez, S.; Barceló, D.; Aga, D.S. Trace Analysis of Polar Pharmaceuticals in Wastewater by LC-MS-MS: Comparison of Membrane Bioreactor and Activated Sludge Systems. *J. Chromatogr. Sci.* **2009**, *47*, 19–25. [CrossRef]
40. Kim, J.-W.; Tue, N.M.; Isobe, T.; Misaki, K.; Takahashi, S.; Viet, P.H.; Tanabe, S. Contamination by perfluorinated compounds in water near waste recycling and disposal sites in Vietnam. *Environ. Monit. Assess.* **2012**, *185*, 2909–2919. [CrossRef]
41. Loos, R.; Wollgast, J.; Huber, T.; Hanke, G. Polar herbicides, pharmaceutical products, perfluorooctanesulfonate (PFOS), perfluorooctanoate (PFOA), and nonylphenol and its carboxylates and ethoxylates in surface and tap waters around Lake Maggiore in Northern Italy. *Anal. Bioanal. Chem.* **2007**, *387*, 1469–1478. [CrossRef]

42. Smeets, J.M.; Rankouhi, T.R.; Nichols, K.M.; Komen, H.; Kaminski, N.E.; Giesy, J.P.; van den Berg, M. In VitroVitellogenin Production by Carp (*Cyprinus carpio*) Hepatocytes as a Screening Method for Determining (Anti)Estrogenic Activity of Xenobiotics. *Toxicol. Appl. Pharmacol.* **1999**, *157*, 68–76. [CrossRef]
43. Cocci, P.; Mosconi, G.; Palermo, F.A. Changes in expression of microRNA potentially targeting key regulators of lipid metabolism in primary gilthead sea bream hepatocytes exposed to phthalates or flame retardants. *Aquat. Toxicol.* **2019**, *209*, 81–90. [CrossRef]
44. Cocci, P.; Mosconi, G.; Palermo, F.A. Pregnane X receptor (PXR) signaling in seabream primary hepatocytes exposed to extracts of seawater samples collected from polycyclic aromatic hydrocarbons (PAHs)-contaminated coastal areas. *Mar. Environ. Res.* **2017**, *130*, 181–186. [CrossRef] [PubMed]
45. Ribecco, C.; Baker, M.E.; Šášik, R.; Zuo, Y.; Hardiman, G.; Carnevali, O. Biological effects of marine contaminated sediments on Sparus aurata juveniles. *Aquat. Toxicol.* **2011**, *104*, 308–316. [CrossRef]
46. Vieira, F.A.; Pinto, P.I.; Guerreiro, P.M.; Power, D.M. Divergent responsiveness of the dentary and vertebral bone to a selective estrogen-receptor modulator (SERM) in the teleost Sparus auratus. *Gen. Comp. Endocrinol.* **2012**, *179*, 421–427. [CrossRef]
47. Cabas, I.; Liarte, S.; García-Alcázar, A.; Meseguer, J.; Mulero, V.; García-Ayala, A. 17α-Ethynylestradiol alters the immune response of the teleost gilthead seabream (Sparus aurata L.) both in vivo and in vitro. *Dev. Comp. Immunol.* **2012**, *36*, 547–556. [CrossRef] [PubMed]
48. Pérez-Sánchez, J.; Borrel, M.; Bermejo-Nogales, A.; Benedito-Palos, L.; Saera-Vila, A.; Calduch-Giner, J.A.; Kaushik, S. Dietary oils mediate cortisol kinetics and the hepatic mRNA expression profile of stress-responsive genes in gilthead sea bream (Sparus aurata) exposed to crowding stress. Implications on energy homeostasis and stress susceptibility. *Comp. Biochem. Physiol. Part D Genom. Proteom.* **2013**, *8*, 123–130. [CrossRef] [PubMed]
49. Navas, J.M.; Segner, H. Vitellogenin synthesis in primary cultures of fish liver cells as endpoint for in vitro screening of the (anti)estrogenic activity of chemical substances. *Aquat. Toxicol.* **2006**, *80*, 1–22. [CrossRef] [PubMed]
50. Gallagher, P.E.; Li, P.; Lenhart, J.R.; Chappell, M.C.; Brosnihan, K.B. Estrogen Regulation of Angiotensin-Converting Enzyme mRNA. *Hypertension* **1999**, *33*, 323–328. [CrossRef]
51. Zilberman, J.M.; Licy, Y.L.; Sartori-Vallinoti, J.C.; Iliescu, R.; Reckelhoff, J.F. Angiotensin converting enzyme inhibitor upreg-ulates the expression of estrogen receptors in the kidney in old female rats. *Faseb J.* **2008**, *22*, 941–948. [CrossRef]
52. Lau, C.; Anitole, K.; Hodes, C.; Lai, D.; Pfahles-Hutchens, A.; Seed, J. Perfluoroalkyl Acids: A Review of Monitoring and Toxicological Findings. *Toxicol. Sci.* **2007**, *99*, 366–394. [CrossRef]
53. DRBC. Contaminants of Emerging Concern in the Delaware River Basin. 2016. Available online: http://www.nj.gov/drbc/quality/reports/emerging/ (accessed on 16 April 2021).
54. Lindstrom, A.B.; Strynar, M.J.; Libelo, E.L. Polyfluorinated Compounds: Past, Present, and Future. *Environ. Sci. Technol.* **2011**, *45*, 7954–7961. [CrossRef] [PubMed]
55. Singh, S.; Singh, S.K. Effect of gestational exposure to perfluorononanoic acid on neonatal mice testes. *J. Appl. Toxicol.* **2019**, *39*, 1663–1671. [CrossRef] [PubMed]
56. Benninghoff, A.D.; Bisson, W.H.; Koch, D.C.; Ehresman, D.J.; Kolluri, S.K.; Williams, D.E. Estrogen-Like Activity of Perfluoroalkyl Acids In Vivo and Interaction with Human and Rainbow Trout Estrogen Receptors In Vitro. *Toxicol. Sci.* **2010**, *120*, 42–58. [CrossRef] [PubMed]
57. Jantzen, C.E.; Annunziato, K.A.; Bugel, S.M.; Cooper, K.R. PFOS, PFNA, and PFOA sub-lethal exposure to embryonic zebrafish have different toxicity profiles in terms of morphometrics, behavior and gene expression. *Aquat. Toxicol.* **2016**, *175*, 160–170. [CrossRef] [PubMed]
58. Jurima-Romet, M.; Huang, H.S.; Paul, C.J.; Thomas, B.H. Enalapril cytotoxicity in primary cultures of rat hepatocytes. II. Role of glutathione. *Toxicol. Lett.* **1991**, *58*, 269–277. [CrossRef]
59. Purclutepe, O.; Iskender, G.; Kiper, H.D.; Tezcanli, B.; Selvi, N.; Avci, C.B.; Kosova, B.; Gokbulut, A.A.; Sahin, F.; Baran, Y.; et al. Enalapril-induced apoptosis of acute promyelocytic leukaemia cells involves STAT5A. *Anticancer. Res.* **2012**, *32*, 2885–2893. [PubMed]
60. Louisse, J.; Rijkers, D.; Stoopen, G.; Janssen, A.; Staats, M.; Hoogenboom, R.; Kersten, S.; Peijnenburg, A. Perfluorooctanoic acid (PFOA), perfluorooctane sulfonic acid (PFOS), and perfluorononanoic acid (PFNA) increase triglyceride levels and decrease cholesterogenic gene expression in human HepaRG liver cells. *Arch. Toxicol.* **2020**, *94*, 3137–3155. [CrossRef]
61. Aerni, H.-R.; Kobler, B.; Rutishauser, B.V.; Wettstein, F.E.; Fischer, R.; Giger, W.; Hungerbühler, A.; Marazuela, M.D.; Peter, A.; Schönenberger, R.; et al. Combined biological and chemical assessment of estrogenic activities in wastewater treatment plant effluents. *Anal. Bioanal. Chem.* **2003**, *378*, 688–696. [CrossRef]
62. Lozano, N.; Rice, C.P.; Pagano, J.; Zintek, L.; Barber, L.B.; Murphy, E.W.; Nettesheim, T.; Minarik, T.; Schoenfuss, H.L. Concentration of organic contaminants in fish and their biological effects in a wastewater-dominated urban stream. *Sci. Total. Environ.* **2012**, *420*, 191–201. [CrossRef]
63. Madsen, L.L.; Korsgaard, B.; Pedersen, K.L.; Bjerregaard, L.B.; Aagaard, T.; Bjerregaard, P. Vitellogenin as biomarker for estrogenicity in flounder Platichthys flesus in the field and exposed to 17α-ethinylestradiol via food and water in the laboratory. *Mar. Environ. Res.* **2013**, *92*, 79–86. [CrossRef] [PubMed]
64. Ankley, G.T.; Jensen, K.M.; Kahl, M.D.; Korte, J.J.; Makynen, E.A. Description and evaluation of a short-term reproduction test with the fathead minnow (Pimephales promelas). *Environ. Toxicol. Chem.* **2001**, *20*, 1276–1290. [CrossRef] [PubMed]

65. Hinck, J.E.; Blazer, V.S.; Denslow, N.D.; Echols, K.R.; Gross, T.S.; May, T.W.; Anderson, P.J.; Coyle, J.J.; Tillitt, D.E. Chemical contaminants, health indicators, and reproductive biomarker responses in fish from the Colorado River and its tributaries. *Sci. Total. Environ.* **2007**, *378*, 376–402. [CrossRef]
66. Pomatto, V.; Palermo, F.; Mosconi, G.; Cottone, E.; Cocci, P.; Nabissi, M.; Borgio, L.; Polzonetti-Magni, A.M.; Franzoni, M.F. Xenoestrogens elicit a modulation of endocannabinoid system and estrogen receptors in 4NP treated goldfish, Carassius auratus. *Gen. Comp. Endocrinol.* **2011**, *174*, 30–35. [CrossRef] [PubMed]
67. Larsen, B.K.; Bjørnstad, A.; Sundt, R.C.; Taban, I.C.; Pampanin, D.M.; Andersen, O.K. Comparison of protein expression in plasma from nonylphenol and bisphenol A-exposed Atlantic cod (*Gadus morhua*) and turbot (*Scophthalmus maximus*) by use of SELDI-TOF. *Aquat. Toxicol.* **2006**, *78*, S25–S33. [CrossRef]
68. Petersen, K.; Tollefsen, K.E. Assessing combined toxicity of estrogen receptor agonists in a primary culture of rainbow trout (Oncorhynchus mykiss) hepatocytes. *Aquat. Toxicol.* **2011**, *101*, 186–195. [CrossRef]
69. Griffin, L.B.; January, K.E.; Ho, K.W.; Cotter, K.A.; Callard, G.V. Morpholino-Mediated Knockdown of ERα, ERβa, and ERβb mRNAs in Zebrafish (*Danio rerio*) Embryos Reveals Differential Regulation of Estrogen-Inducible Genes. *Endocrinology* **2013**, *154*, 4158–4169. [CrossRef] [PubMed]
70. Nelson, E.R.; Habibi, H.R. Functional Significance of Nuclear Estrogen Receptor Subtypes in the Liver of Goldfish. *Endocrinology* **2010**, *151*, 1668–1676. [CrossRef]
71. Murakami, M.; Imamura, E.; Shinohara, H.; Kiri, K.; Muramatsu, Y.; Harada, A.; Takada, H. Occurrence and Sources of Perfluorinated Surfactants in Rivers in Japan. *Environ. Sci. Technol.* **2008**, *42*, 6566–6572. [CrossRef]

Article

Assessment of the Levels of Pollution and of Their Risks by Radioactivity and Trace Metals on Marine Edible Fish and Crustaceans at the Bay of Bengal (Chattogram, Bangladesh)

Krishna Prasad Biswas [1], Shahadat Hossain [2,3], Nipa Deb [2], A.K.M. Saiful Islam Bhuian [2,*], Sílvia C. Gonçalves [4,*], Shahadat Hossain [2] and Mohammad Belal Hossen [1]

1 Department of Physics, Chittagong University of Engineering and Technology, Chattogram 4349, Bangladesh; kpbcu@yahoo.com (K.P.B.); belalcuet@gmail.com (M.B.H.)
2 Atomic Energy Centre, Bangladesh Atomic Energy Commission, Chattogram 4209, Bangladesh; sahedmc@gmail.com (S.H.); nipaacdu@yahoo.com (N.D.); shahadat.baec@gmail.com (S.H.)
3 Interdisciplinary Graduate School of Engineering Sciences, Kyushu University, Fukuoka 816-8580, Japan
4 MARE—Marine and Environmental Sciences Centre, ESTM-School of Tourism and Maritime Technology, Polytechnic of Leiria, 2520-641 Peniche, Portugal
* Correspondence: saiful.bhuian@baec.gov.bd (A.S.I.B.); silvia.goncalves@ipleiria.pt (S.C.G.); Tel.: +88-0199-403-4539 (A.S.I.B.)

Citation: Biswas, K.P.; Hossain, S.; Deb, N.; Bhuian, A.K.M.S.I.; Gonçalves, S.C.; Hossain, S.; Hossen, M.B. Assessment of the Levels of Pollution and of Their Risks by Radioactivity and Trace Metals on Marine Edible Fish and Crustaceans at the Bay of Bengal (Chattogram, Bangladesh). *Environments* **2021**, *8*, 13. https://doi.org/10.3390/environments8020013

Academic Editors: Claude Fortin and Vernon Hodge

Received: 10 November 2020
Accepted: 1 February 2021
Published: 11 February 2021

Abstract: Marine environmental pollution is a longstanding global problem and has a particular impact on the Bay of Bengal. Effluent from different sources directly enters rivers of the region and eventually flows into the Bay of Bengal. This effluent may contain radioactive materials and trace metals and pose a serious threat to the coastal environment, in addition to aquatic ecosystems. Using gamma spectrometry and atomic absorption spectrometry, a comprehensive study was carried out on the radioactivity (^{226}Ra, ^{232}Th, ^{40}K, and ^{137}Cs) and trace metal (Cd, Pb, Zn, Cu, Ni, Fe, Mn, and Cr) concentrations, respectively, in fish and crustacean species collected from the coastal belt of the Bay of Bengal (Chattogram, Bangladesh). The analysis showed a noticeable increment in the levels of different radioactive pollutants in the marine samples, although the consumption of the studied fish and crustacean species should be considered safe for human health. Anthropogenic radionuclide (^{137}Cs) was not detected in any sample. Furthermore, the metal concentrations of a small number of trace elements (Pb, Cd, Cr) were found to be higher in most of the samples, which indicates aquatic fauna are subject to pollution. The estimated daily intake (EDI), target hazard quotient (THQ), hazard index (HI), and target cancer risk (TR) were calculated and compared with the permissible safety limits. It was found that consuming the seafood from the Bay of Bengal may cause adverse health impacts if consumption and/or means of pollution are not controlled.

Keywords: radioactive materials; trace metals; bioaccumulation; marine fish; crustaceans; marine environmental pollution; Bay of Bengal

1. Introduction

The marine environment is one of the most important sources of life on Earth and performs a number of key environmental functions for the lives and livelihood of humans and other organisms. Marine environmental pollution is a longstanding global problem. Marine pollution affects the Bay of Bengal because of the vulnerability of its aquatic habitats to such pollution following the industrial revolution in the 19th century. Most aquatic ecosystems can cope with a certain degree of pollution, but severe pollution is reflected in changes in the fauna and flora of their communities [1]. Developing countries, including Bangladesh, are most affected by this human-made problem. Bangladesh is considered to be one of the most suitable countries for the Blue Economy following the recent establishment of a vast maritime boundary in the Bay of Bengal. The result of two verdicts on maritime boundaries, also involving Myanmar and India, allows Bangladesh to exclusively exercise its sovereign rights over 118,813 sq km of water. The affected area extends up to 12 nautical miles into territorial sea and includes a further Exclusive Economic Zone (EEZ) of 200 nautical miles [2]. However, agriculture runoff, untreated sewage, and industrial pollution are reportedly being discharged directly into rivers, eventually flowing into the Bay of Bengal. This effluent may contain radioactive materials and trace metals and pose a particularly serious threat to the coastal environment and the aquatic ecosystems [3,4].

The main causes of radioactivity in the marine environment are seabed movement originating from underwater volcanic activity and undersea earthquakes, natural processes of weathering, and mineral recycling of terrestrial rocks [5]. Different types of geological materials and minerals, such as ores and igneous rocks, which often contain large concentrations of natural radioactive elements, contribute to the transfer of radionuclides to water through leaching action [3]. The level of naturally occurring radioactive elements in the marine environment has gradually increased due to human activities, such as mining and processing of ores, production of natural oil and gas, and combustion of fossil fuels in coal-fired power plants. Moreover, anthropogenic contributions are made by underwater nuclear device tests, post-nuclear disposal of industrial and radioactive waste, recycling of spent nuclear fuel, and accidents, including leaks from nuclear power plants [4–7]. Among the radionuclides, uranium, radium, and radon are soluble in seawater, whereas thorium is almost completely insoluble. These can dissolve in seawater and then attach to sediment on the seabed and suspended plankton in the seawater. These dissolved radionuclides and plankton, in the long run, contaminate marine organisms, including fish, crustaceans, and several types of shellfish [8].

In addition, trace metals are released into the environment from different sources, such as transportation, industrial activities, fossil fuels, agriculture, urbanization, and other human activities [6–9]. The release of large quantities of trace metals into nature has resulted in several environmental problems due to their non-biodegradability and persistence. Marine life can have a considerable aptitude for bioaccumulation and biosorption of trace metals [10–12]. Under certain environmental conditions, such as consolidated effects of biotic and abiotic factors, such as plants, animals, and microbes; the amount of sunlight in the ecosystem; the amount of oxygen and nutrients dissolved in the water; proximity to land; depth; and temperature; these trace metals may accumulate to a toxic concentration and cause ecological damage by threatening the health of aquatic and terrestrial organisms, including humans [1,13,14]. The principal pathway leading to human exposure to radionuclides and trace metals is the consumption of seafood. A large variety of seafood is eaten. Fish and crustacean species, because of their important dual role as both an integral element of marine food webs and a commercial food product for humans, are frequently selected as indicator organisms in studies related to pollution or the bioaccumulation processes of various noxious substances [15]. The bioaccumulation of trace metals and isotopic contaminants by the tissues and organs of marine organisms has been studied globally and has led to the adoption of the bioindicator concept for environmental quality assessments [5–8,16,17].

Research on the bioaccumulation of radionuclides and trace metals in the most commonly consumed fish and crustacean species in the Bay of Bengal region is limited. Given the importance of such knowledge, this study was carried out to determine the radioactivity and trace metal levels in six marine fish and four crustacean species of the northern coastal belt of the Bay of Bengal, Bangladesh. The bioaccumulation levels of radioactive materials (^{226}Ra, ^{232}Th, ^{40}K, and ^{137}Cs) and trace metals (Cd, Pb, Zn, Cu, Ni, Fe, Mn, and Cr) were estimated by gamma spectrometry and atomic absorption spectrometry, respectively. Different radiological and health-hazard parameters were also calculated to assess the marine environmental quality in this region concerning these pollutants, in addition to the health risks resulting from the consumption of the studied seafood.

2. Materials and Methods

2.1. Sample Collection and Preparation

The studied field locations were openly accessible, and the organisms collected did not belong to any protected species, so permissions were not required at the time of collecting samples from the studied locations. The investigated fish and crustaceans were obtained from four fish/seafood markets of the northern coastal belt of the Bay of Bengal, Chattogram, Bangladesh (Figure 1), during the rainy (July) and autumn (September and October) seasons of 2017. Bangladesh is a country of six seasons (summer, rainy, fall, autumn, winter, and spring). The fish investigated in this study are mostly available in rainy and autumn seasons. Hence, we obtained our samples in these two seasons. The fish samples were obtained from fishing boats operating in the corresponding locations, while the crustacean samples were obtained from a local sea food market in Chattogram city. A total of 20 samples (10 samples from each season) were chosen for investigation. The studied organisms were six fish and four crustacean species, locally available and commercially important, i.e., *Tenualosa ilisha* (Ilish), *Harpodon neherues* (Lotia), *Sillaginopsis panijus* (Sundora Baila), *Sardinella longiceps* (Colombo), *Trichiurus lepturus* (Churi), and *Konosirus punctatus* (Shad) concerning the fish species, and *Scylla serrata* (mud crab), *Portunus sanguinolentus* (three-spot swimming crab), *Hemigrapsus takanoi* (Asian shore crab) and *Penaeus semisulcatus* (cat tiger shrimp) concerning the crustaceans.

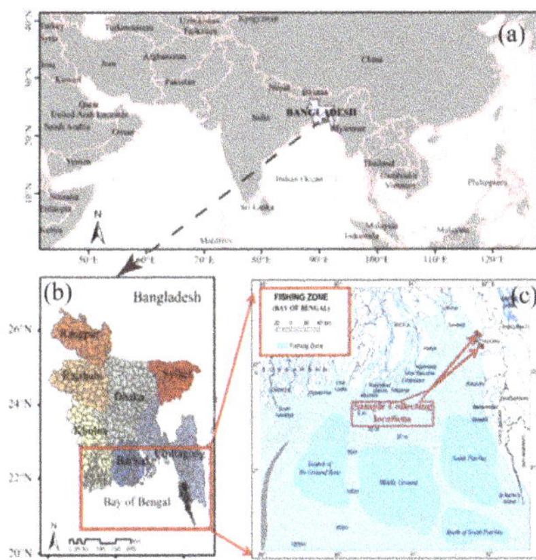

Figure 1. Location of the study area on the (**a**) world map, (**b**) Bangladesh map, and (**c**) Bay of Bengal map.

The details of the collected samples are given in Table 1. All the samples were labelled, stored in ice, and, on the same day, transported to the laboratory and washed with clean water and distilled water. Afterwards, the samples were dried in filter paper, packed in polyethylene bags, and stored in a refrigerator of the Laboratory of Atomic Energy Centre, Chattogram, to avoid degradation, spoiling, contamination, or any other decomposition until further treatment and analysis were done. The collected fish and crustacean samples were measured for the total wet weight (WW) and recorded beforehand. For the analysis of radionuclides and trace metal concentrations, all the animals were used in their entirety, since the analysis was intended to quantify the radioactivity and the metal concentrations of each organism.

Table 1. Marine organisms collected and analyzed during rainy and autumn seasons of 2017 at the Bay of Bengal (Bangladesh).

Sample Type	Name of the Organism/Local Name	Scientific Name	Range of the Weight in gm	Range of the Length in cm	Average Moisture	Total Sample Weight in Kg	Sampling Location and Season
6 Fish samples	Hilsa fish/ Ilish fish	*Tenualosa ilisha*	588–626	38.0–38.5	86.01%	1.214	(L-1) and R
			274–306	29.0–30.0		1.204	(L-2) and A
	Bombay duck/ Lotia fish	*Harpodon neherues*	62–102	22.0–25.5	90.22%	0.844	(L-1) and R
			22–88	18.0–27.0		0.890	(L-2) and A
	Flathead sillago/ Sundara Baila	*Sillaginopsis panijus*	82–150	25.0–29.0	83.01%	0.960	(L-1) and R
			168–320	30.0–38.0		1.152	(L-1) and A
	Indian oil sardine/ Colombo fish	*Sardinella longiceps*	50–68	18.5–19.5	77.85%	1.014	(L-1) and R
			42–62	18.5–20.5		1.176	(L-1) and A
	Belt fish/ Churi Fish	*Trichiurus lepturus*	84–106	44.0–51.0	80.88%	0.964	(L-1) and R
			240–322	62.0–73.0		1.116	(L-1) and A
	Dotted gizzard shad/Shard fish	*Konosirus punctatus*	270–292	22.0–25.0	73.41%	0.850	(L-1) and R
			255–285	21.0–23.0		0.820	(L-1) and A
4 Crustacean samples	Three-spot swimming crab	*Portunus san-guinolentus*	90–140	-	76.44%	1.452	(L-3) and R
			93–145	-		1.485	(L-3) and A
	Mud crab	*Scylla serrata*	60–120	-	82.82%	1.441	(L-4) and R
			55–120	-		1.411	(L-4) and A
	Asian Shore crab/ Chati Kakra	*Hemigrapsus takanoi*	8–18	-	78.00%	1.438	(L-3) and R
			8–20	-		1.410	(L-3) and A
	Cat tiger shrimp/ Harina Chingri	*Penaeus semisulcatus*	12–24	12.0–14.5	76.51%	0.880	(L-1) and R
			10–26	12.0–16.0		0.850	(L-1) and A

Note: (L-1)—Fishery Ghat, Chattogram; (L-2)—Gohira Fishery Ghat, Anwara, Chattogram; (L-3)—Ganga Bari Fish Market, Chattogram; (L-4)—Hazari Goli Seafood Market, Chattogram; R—rainy season and A—autumn season.

Each sample was sun-dried at least one week to remove the extra water and subsequently oven-dried at about 70 °C to obtain a constant weight. The samples were then re-weighed to determine the dry/wet ratio. Finally, the dried samples were ground, sieved for homogeneity, and stored in clean and dry uncontaminated empty cylindrical plastic containers of uniform size (2.8 cm × 8.0 cm), sealed with wide vinyl adhesive tapes around their screw necks to air tighten for succeeding uses.

For the trace metal determination, a part of the powdered samples was kept aside in clean and dry sealable plastic bags and kept in an air-tight glass jar to save the samples from any types of chemical reactions. Five grams of the homogenized powder were taken from each specimen and placed in a 250 mL digestion beaker. A digestion mixture containing 6.0 mL of high purity nitric acid (70%) and 2.0 mL of hydrochloric acid (37%) were added and heated at 60 °C for half an hour. After cooling down for 15 min, 4.0 mL of hydrogen peroxide (30%) was added to each beaker and heated until a clear solution was obtained and the volume decreased into half of its original volume following the standard operating procedure described elsewhere [18]. The digested portions were filtered through Whatman filter paper (No. 42) and diluted to a final volume of 50 mL using de-ionized water.

2.2. Methods for Determining Radioactivity and Trace Metals

2.2.1. Gamma Spectrometry Analysis

For the natural and artificial radioactivity measurement of each sample, about 100 g of dried fish/crustacean samples were weighed and placed in individual plastic containers. The samples were kept 4 weeks before the analysis in airtight conditions to allow secular equilibrium between thorium and radium and their short-lived progenies [14].

The activity concentrations of ^{226}Ra, ^{232}Th, ^{40}K, and ^{137}Cs in the samples were determined using an NATS GCD-40 190 p-type HPGe (Baltic Scientific Instruments, Riga, Latvia), gamma-ray spectrometer (63.3 mm crystal diameter, 61.3 mm thickness, +2200 V operating bias voltage). The low background detector of relative efficiency of 43.1% and an energy resolution of 1.74 keV full width at half maximum (FWHM) at the 1332 keV peak of ^{60}Co was enclosed in a cylindrical lead shield. The counting system was connected to an 8 k multi-channel analyzer (MCA 527, GBS Elektronik GmbH) with associated electronics for data acquisition of photo-peak areas. A Spectral Line GP© was used to analyze the gamma-ray counts received from the samples. Energy calibration and the absolute photo-peak efficiency evaluation were performed using standard reference material (Code: 8501-EG-SVE, Eckert and Ziegler Analytics), diluted with a multi-nuclide gamma-ray source (^{241}Am, ^{109}Cd, ^{57}Co, ^{139}Ce, ^{113}Sn, ^{85}Sr, ^{137}Cs, ^{88}Y, ^{60}Co) having homogeneously distributed activity, and maintaining the same geometry and density as the plastic containers containing the samples. To minimize the statistical counting error, the samples were counted for a period of 30,000 s. An empty container was also counted under the same conditions to determine the background counts. To calculate the specific activities, the background counts for the same counting condition were deducted from the counts of each sample to obtain the net counts. For spectrum analyses, the single transition gamma-ray line 1460.822 keV was used to determine the activity concentrations of ^{40}K. The gamma-ray photo-peaks of 295.221 keV and 351.922 keV from ^{214}Pb, and 609.320 keV, 1120.310 keV and 1764.551 keV from ^{214}Bi were used to determine the activity concentrations of ^{226}Ra. The activity concentrations of ^{232}Th were determined using the net counts under the 238.630 keV and 300.087 keV photo-peaks from ^{212}Pb, 911.205 keV and 968.970 keV photo peaks from ^{228}Ac, and 583.190 keV and 2614.533 keV from ^{208}Tl. For the evaluation of ^{226}Ra and ^{232}Th activity, a weighted mean approach was applied using the aforementioned gamma lines [19,20]. Much care was taken to prevent contamination during the investigation.

2.2.2. Determination of Trace Metals

The fish and crustacean samples were analyzed for eight trace elements, Cd, Pb, Zn, Cu, Ni, Fe, Mn, and Cr, using an atomic absorption spectrophotometer (HITACHI Z-2000) [21]. Four standard solutions of different known concentrations were prepared, and the elemental concentrations in the unknown samples were determined by extrapolation from the calibration curve. The digested samples were directly aspirated into the flame (air–acetylene fuel mixture). The concentration corresponding to the absorption in the digest was determined by using the absorption mode. The minimum detection limits of the analyzer for the investigated trace metals in ppm are 0.002 (Cd), 0.010 (Cu), 0.005 (Zn), 0.050 (Pb), 0.010 (Mn), 0.020 (Fe), 0.020 (Ni), and 0.020 (Cr) [22]. The limits of detection (LODs) for all the elements analyzed in the samples were calculated as the blank signal plus three times its standard deviation, whereas the limit of quantification (LOQ) was calculated as ten times the standard deviation of the blank signal, following [23,24]. The recovery estimation was performed by the laboratory during method development and validation processes using standard reference materials (SRMs) and certified reference materials (CRMs). However, this time we used secondary reference material made from commercial standard solutions of each element. At each step of the digestion processes, acid blanks (laboratory blank) were prepared to ensure that the samples and chemicals used were not contaminated. Each set of digestion had its acid blank and was corrected by using its blank. The measurement of each sample was taken three times. All the experimental values were reported as mean value ± standard deviation (SD). Throughout the analysis, all the trace metal standards were prepared and run to check the precision of the instrument. The quality control and quality assurance protocol set by the U.S. Environmental Protection Agency for metal analysis were used. The quality assurance testing relied on the blank controls, and the yield of the chemical procedure was used.

2.3. Radiation Dose and Risk Assessment

2.3.1. Radiation Dose Assessment

Activity Concentration

The activity concentrations (Bq kg^{-1}) of the natural and anthropogenic radionuclides in the measured samples were evaluated by the following equation [25]:

$$A\left(Bq\ kg^{-1}\right) = \frac{CPS \times 1000}{\varepsilon_{\gamma} I_{\gamma} M} \quad (1)$$

where CPS = net count per second (i.e., CPS for sample—CPS for background), ε_{γ} = detector efficiency of the specific γ-ray, I_{γ} = intensity of the gamma-ray, and, M = mass of the sample in grams.

The errors of the measurements were expressed in terms of the standard deviation of the $\pm 1\sigma$ level.

Total Effective Dose

Annual effective dose is a useful concept that enables the radiation doses from different radionuclides and from different types of sources of radioactivity to be added. Estimation of the radiation-induced health effects associated with the intake of radionuclides in the body is proportional to the total dose delivered by the radionuclides while resident in the various organs. Radiation doses ingested are obtained by measuring radionuclide activity in foodstuffs ($Bqkg^{-1}$) and multiplying these by the masses of food consumed over a certain period (kgd^{-1} or kgy^{-1}). A dose conversion factor ($SvBq^{-1}$) can then be applied to give an estimate of the ingestion dose. Thus, the ingested dose is given by [26,27]:

$$\text{annual effective dose}\left(Svy^{-1}\right) = \text{concentration}\left(Bq.kg^{-1}\right) \times \text{annual intake}\left(kgy^{-1}\right) \times DCF\left(SvBq^{-1}\right) \quad (2)$$

where DCF is the standard dose conversion factor, which is equal to 0.2800 $\mu SvBq^{-1}$ for ^{226}Ra, 0.2300 $\mu SvBq^{-1}$ for ^{232}Th, and 0.0062 $\mu SvBq^{-1}$ for ^{40}K [25].

Therefore, the total effective dose (Svy^{-1}) via ingestion is calculated by the following formula:

$$\text{total annual effective dose} = \left\{C_R(^{226}Ra) \times I_F \times E_D\right\} + \left\{C_R(^{232}Th) \times I_F \times E_D\right\} + \left\{C_R(^{40}K) \times I_F \times E_D\right\} \quad (3)$$

where C_R is the concentration of radionuclides in ingested fish/crustaceans ($Bqkg^{-1}$), I_F is the annual intake (kgy^{-1}) of fish/crustaceans containing radionuclides, and E_D is the ingestion dose conversion factor for radionuclides ($SvBq^{-1}$). The intake rates for Bangladeshi consumers were taken from the "Year Book of Fisheries Statistics of Bangladesh 2018" [28].

2.3.2. Radiation Risk Assessment

Internal Hazard Index (H_{in})

Radon and its short-lived descendants are hazardous to the respiratory organs. The internal hazard index (H_{in}) is used to quantify the internal exposure to radon and its progenies, which is given by the equation:

$$H_{in} = \frac{C_{Ra}}{185} + \frac{C_{Th}}{259} + \frac{C_K}{4810} \quad (4)$$

where C_{Ra}, C_{Th}, and C_K are the concentration ($Bqkg^{-1}$) of Ra, Th, and K, respectively.

The values of the internal hazard index (H_{in}) must be less than unity for the radiation hazard to be negligible [29,30].

Excess Lifetime Cancer Risk (ELCR)

Nowadays, cancer is called a life-threatening disease, and the percentage of this disease increases all over the world, including in Bangladesh, due to various reasons. One of the reasons is the effect of radiation on the biological cells, which contributes to a greater extent to increasing cancer incidence. An effort was made to assess the excess lifetime cancer risk due to the ingestion of marine fish and crustaceans by the procedure proposed by the United States Environmental Protection Agency (USEPA) [31]. The following equation was used to calculate the excess lifetime cancer risk [12,32]:

$$\mathrm{ELCR} = \mathrm{A_{ir}} \times \mathrm{A_{ls}} \times \mathrm{R_c} \tag{5}$$

where $ELCR$, A_{ir}, A_{ls}, and R_c are the excess lifetime cancer risk, the annual intake of radionuclide (Bq), the average lifespan 72 yr. (life expectancy of a male and a female are 70.6 years and 73.5 years, respectively in Bangladesh [33]), and the mortality cancer risk coefficient (Bq^{-1}), respectively. The values of mortality cancer risk coefficients are 9.56×10^{-9} Bq^{-1} for ^{226}Ra, 2.45×10^{-9} Bq^{-1} for ^{232}Th, and 5.89×10^{-10} Bq^{-1} for ^{40}K [12,31]. The acceptable ELCR limit is 10^{-3} for radiological risk in general [32,34].

2.4. Health Risk Assessment of Trace Metals

2.4.1. Estimated Daily Intake (EDI)

EDI is measured by the following equation in (mgkg^{-1} body weight per day) [35]:

$$\mathrm{EDI} = \frac{\mathrm{E_F} \times \mathrm{E_D} \times \mathrm{F_{IR}} \times \mathrm{C_f} \times \mathrm{C_M}}{\mathrm{W_{AB}} \times \mathrm{AT_n}} \times 10^{-3} \tag{6}$$

where E_F is the exposure frequency (365 days per year), E_D=72 years is the exposure duration, F_{IR} is the ingestion rate, which is taken as 62.58 g per person per day [28], C_f is the conversion factor (C_f = 0.208) to convert fresh weight to dry weight considering 79% of moisture content of the fish fillet [35,36], C_M is the metal concentration in fish fillet (mgkg^{-1} dry weight basis), W_{AB} is the average body weight (60 kg) of Bangladeshi adult people [37], and AT_n (equal to $E_F \times E_D$) is the average exposure time for non-carcinogens [35].

Several organizations such as WHO, FAO, etc., provided guidelines on the intake of metals by a human. The maximum tolerable daily intake (MTDI), provisional tolerable daily intake (PTDI), and the acceptable daily intake (ADI) are used to describe the safe levels of intake for several toxins, including toxic metals. The EDI of trace metals measures the amount of metal intake by an adult (60 kg) by ingestion of fish or crustaceans per day.

2.4.2. Target Hazard Quotient (THQ)

The THQ is non-carcinogenic risk and is dimensionless. In this study, the non-carcinogenic health risks associated with the consumption of fish and crustacean species were assessed based on the target hazard quotients (THQs), and their calculations were done using the standard assumption for an integrated USEPA risk analysis as follows [38]:

$$\mathrm{THQ} = \frac{\mathrm{E_F} \times \mathrm{E_D} \times \mathrm{F_{IR}} \times \mathrm{C_f} \times \mathrm{C_M}}{\mathrm{W_{AB}} \times \mathrm{AT_n} \times \mathrm{RfD}} \times 10^{-3} \tag{7}$$

E_F, E_D, F_{IR}, C_f, C_M, W_{AB}, and AT_n are explained in the earlier section. RfD is the reference dose of individual metal (mg/kg/day) (Table 2). The RfD represents an estimate of the daily exposure to which the human population may be continually exposed over a lifetime without a significant risk of deleterious effects. If the THQ is less than 1, the exposed population is unlikely to experience obvious adverse effects. If the THQ is equal to or higher than 1, then there is a potential health risk [39], and related interventions and protective measurements should be taken.

Table 2. Reference dose (RfD) and carcinogenic slope factor (CSFo), oral.

Trace Elements	RfD(Mg/Kg/Day)	CSFo(Mg/Kg bw/Day)$^{-1}$	Reference of CSFo
Pb	4×10^{-3}	8.5×10^{-3}	[37]
Cu	4×10^{-2}	-	[37]
Zn	3×10^{-1}	-	[40]
Mn	1.4×10^{-1}	-	[40]
Cd	1×10^{-3}	1.5×10^{1}	[41]
Ni	2×10^{-2}	1.7×10^{0}	[40]
Cr	3×10^{-3}	5×10^{-1}	[41]
Fe	7×10^{-1}	-	[40]

2.4.3. Hazard Index (HI)

To estimate the overall potential health risk related with more than one metal, THQ of every metal is summed up and referred to as hazard index (HI). The HI can be calculated by the sum of the target hazard quotients (THQs) of each metal:

$$HI = THQMn + THQFe + THQCu + THQZn + THQPb + THQCd + THQCr + THQNi \quad (8)$$

2.4.4. Target Cancer Risk

For carcinogens, the target cancer risk (lifetime cancer risk) is estimated as the incremental probability of an individual to develop cancer over a lifetime exposure to that potential carcinogen (i.e., incremental or excess individual lifetime cancer risk) [38]. Acceptable risk levels for carcinogens range from 10^{-4} (risk of developing cancer over a human lifetime is 1 in 10,000) to 10^{-6} (risk of developing cancer over a human lifetime is 1 in 1,000,000). The equation used for estimating the target cancer risk, which is dimensionless, is as follows [38]:

$$TR = \frac{E_F \times E_D \times F_{IR} \times C_f \times C_M \times CSF_o}{W_{AB} \times AT_C} \times 10^{-3} \quad (9)$$

Here, CSFo is the oral carcinogenic slope factor (mg/kg BW/day)$^{-1}$, and ATc (equal to $E_F \times E_D$) is the average exposure time for carcinogens. Since Mn, Fe, Cu, and Zn do not cause any carcinogenic effects, their CSFo have yet not been established in USEPA 2012. Thus, TR values for the intake of Pb, Cd, Cr, and Ni were calculated to show the carcinogenic risk using oral carcinogenic slope factors (CSFo) of these toxic elements given in Table 2.

3. Results and Discussion

Chattogram is the second largest city in Bangladesh, where the main seaport of Bangladesh is located on the bank of the Karnaphuli river. The Chattogram seaport handles ninety per cent of Bangladesh's export–import trade and has been used by India, Nepal, and Bhutan to join for transshipment. There are thousands of industries including a textile mill, cement factory, tannery, oil refinery, dichlorodiphenyltrichloroethane (DDT) plant, chemical factory, fertilizer factory, paper mills, power plant, dry-dock, paint factory, rayon mills, etc., situated near to the bank of the Karnaphuli river, among which about 200 are identified as pollution causing units continuously discharging unlawfully a huge amount of pollutants that finally enter into the Bay of Bengal [42]. The wastes coming from the municipal sewage system through many canals of Chattogram city is another potential factor of the pollution problem of this bay. Moreover, illegal and/or accidental discharges of grease, fish oil, bilge, garbage, etc., from the merchant and fishing vessels are causing pollution to the Bay of Bengal [43].

3.1. Radiation Dose and Risk Assessment by Radioactivity Analysis

Marine fish and crustaceans are potential bio-indicators when they accumulate the target radionuclides from surrounding waters [15,44]. Monitoring radionuclide levels in fish and crustaceans is of great importance due to their significant contribution to the natural radiation dose received by human beings consuming them [45].

Among all the fish and crustacean samples, ^{226}Ra was detected only in *Harpodon neherues*, and its value was 5 ± 2 Bqkg^{-1}. There was no anthropogenic radionuclide (^{137}Cs) in any sample. In all the studied fish samples, mean activity concentrations of ^{232}Th and ^{40}K ranged from 7 ± 1 to 190 ± 10 and 210 ± 50 to 360 ± 40 Bqkg^{-1}, respectively. Similarly, in crustacean samples, the activity concentrations of ^{232}Th and ^{40}K ranged from 5.0 ± 2 to 53 ± 10 and 130 ± 40 to 240 ± 70 Bqkg^{-1}, respectively. Further, note that the activity concentrations of ^{40}K were greater than those of the other radionuclides for all samples (Table 3). Mean values of the activity concentrations of ^{226}Ra, ^{232}Th, and ^{40}K in all the fish and crustacean samples were 5 ± 2, 67 ± 9, and 250 ± 50 Bqkg^{-1}, respectively.

Table 3. Mean activity concentration (Bqkg^{-1}) and excess lifetime cancer risk (ELCR) due to the consumption of natural radionuclide from marine fish and crustacean samples collected at the Bay of Bengal.

Sl. No.	Types of Samples	Name of Organism/Local Name	Scientific Name	No. of Samples	Mean Activity for ^{226}Ra (Bqkg^{-1})	Mean Activity for ^{232}Th (Bqkg^{-1})	Mean Activity for ^{40}K (Bqkg^{-1})	ELCR for ^{226}Ra	ELCR for ^{232}Th	ELCR for ^{40}K
1	Fish	Hilsa fish/ Ilish fish	*Tenualosa ilisha*	2	BDL	7 ± 1	310 ± 30	BDL	2.60 × 10^{-5}	2.96× 10^{-4}
2		Bombay duck/ Lotia fish	*Harpodon neherues*	2	5 ± 2	170 ± 30	270 ± 60	7.07 × 10^{-5}	6.87 × 10^{-4}	2.62× 10^{-4}
3		Flathead sillago/ Sundara Baila	*Sillaginopsis panijus*	2	BDL	190 ± 10	300 ± 60	BDL	7.60 × 10^{-4}	2.86 × 10^{-4}
4		Indian oil sardine/ Colombo fish	*Sardinella longiceps*	2	BDL	BDL	210 ± 50	BDL	BDL	2.03 × 10^{-4}
5		Belt fish/ Churi fish	*Trichiurus lepturus*	2	BDL	11 ± 2	360 ± 40	BDL	4.23 × 10^{-5}	3.44 × 10^{-4}
6		Dotted gizzard shad/ Shard fish	*Konosirus punctatus*	2	BDL	BDL	250 ± 70	BDL	BDL	2.37 × 10^{-4}
7	Crustacean	Three-spot swimming crab	*Portunus sanguino-lentus*	2	BDL	BDL	190 ± 30	BDL	BDL	1.79 × 10^{-6}
8		Mud crab	*Scylla serrata*	2	BDL	53 ± 10	240 ± 70	BDL	2.12 × 10^{-4}	2.32 × 10^{-4}
9		Asian shore crab/ Chati Kakra	*Hemigrapsus takanoi*	2	BDL	37 ± 6	220 ± 50	BDL	1.47 × 10^{-4}	2.08 × 10^{-4}
10		Cat tiger shrimp/ Harina Chingri	*Penaeus semisulcatus*	2	BDL	5 ± 2	130 ± 40	BDL	2.01 × 10^{-5}	1.30 × 10^{-4}

Note: BDL—below detection limit.

The comparative data of the present study with the previously reported studies of different radioactive elements in fish and crustacean samples of the Bay of Bengal region are reported in Table 4. It was reported in 2012 that the activity concentrations of ^{226}Ra, ^{232}Th, and ^{40}K in sea-fish samples of the Black Sea region of Turkey were 0.06 ± 0.01 to 0.96 ± 0.36, 0.12 ± 0.04 to 1.03 ± 0.15, and 35.04 ± 0.24 to 127.4 ± 2.3 Bqkg^{-1}, respectively [14].

In all these cases, the results were very much lower than those of the present observations. Conversely, the mean activity concentration of ^{226}Ra, ^{232}Th, and ^{40}K in marine fish of the Straits of Malacca was reported as 4.05 ± 0.48 to 7.83 ± 0.78, 1.93 ± 0.24 to 6.21 ± 0.53, and 288 ± 15 to 399 ± 20 Bqkg^{-1}, respectively, in 2015 [12], being considerably comparable to the present study except the concentration of ^{232}Th. The obtained activity concentration of ^{232}Th and ^{40}K in all the fish and crustacean samples were 67 ± 9 and 250 ± 50 Bqkg^{-1}, respectively, which are almost similar to the activity concentration obtained in crab samples from river Odeomi, Ogun State, at the Southwest of Nigeria [46]. It should be noted that if we liken the present study to previous studies for fish [47,48] at the same region, there is a noticeable increment in the levels of different radioactive elements in those marine organisms of the Bay of Bengal region (shown in Table 4). This increment might be due to the wide variety of activities (e.g., housing, tourism, power generation plants, petroleum, steel, shipbuilding, chemical, pharmaceutical, textile, vegetable oil refineries, glass manufacturing industries, etc.) around the Bay of Bengal region, which have been diversified over the years. Multi-functionality and diversification of industries, fisheries, and agriculture are probably linked with these changes in the marine environment. The absence of the anthropogenic radionuclide (^{137}Cs) indicates that there was not any effect on the marine organisms analyzed due to the post-nuclear disposal of industrial and radioactive waste, underwater nuclear device tests, accidents including leaks from nuclear power plants and from recycling of spent nuclear fuel, etc. As Bangladesh is establishing its first nuclear power plant (NPP) at Rooppur, Pabna, this baseline data can help to further monitor these activities.

Table 4. Comparative data of the radioactivity (Bqkg^{-1}) in marine fish and crustacean samples of the Bay of Bengal, Bangladesh.

Sl. No.	Study Year	Number of Species	^{226}Ra	^{232}Th	^{40}K	^{137}Cs	^{238}U	^{228}Ra	Reference
Fish									
1	1995	15	-	0.31 ± 0.05 to 1.67 ± 0.48	8.5 ± 1.2 to 57.1 ± 5.3	BDL to 1.98 ± 0.33	0.31 ± 0.05 to 1.19 ± 0.17	-	M. N. Alam [47]
2	2000	15	0.10 ± 0.03 to 1.66 ± 0.24	-	18.1 ± 3.4 to 86.4 ± 6.7	0.19 ± 0.04 to 1.47 ± 0.28	-	0.39 ± 0.07 to 1.35 ± 0.19	S. Ghose [48]
3	2017	2	-	8.5 ± 9.6 to 13 ± 17	265 ± 417 to 460 ± 310	-	9 ± 19 to 13 ± 14	-	M. H. Kabir [49]
4	2020	6	BDL to 5 ± 2	BDL to 190 ± 10	210 ± 50 to 360 ± 40	BDL	-	-	**Present Study**
Crustacean									
1	1995	5	-	0.36 ± 0.10 to 0.78 ± 0.23	7.32 ± 0.88 to 16.7 ± 1.8	BDL to 0.47 ± 0.07	0.11 ± 0.01 to 0.49 ± 0.18	-	M. N. Alam [47]
2	2020	4	BDL	BDL to 53 ± 10	130 ± 40 to 240 ± 70	BDL	BDL	-	**Present Study**

Note: BDL—below detection limit.

The ELCR values of ^{226}Ra, ^{232}Th, and ^{40}K in all the fish and crustaceans are shown in Table 3. It is to be mentioned here that ELCR for ^{226}R was below detection level (BDL), except for in *Harpodon neherues*. However, all of the obtained values (shown in Table 3) were lower than the acceptable ELCR limit of 10^{-3} for radiological risk in general, indicating no health hazard to the consumers [32,34].

The estimated annual effective doses due to intake of ^{226}Ra, ^{232}Th, and ^{40}K from the consumption of seafood (fish and crabs) from the Bay of Bengal are presented in Table 5, and the comparison between the mean annual effective dose and its world average values are shown in Figure 2. The average annual effective dose for ^{226}Ra and ^{40}K were found to be within UNSCEAR acceptable limits. Though the average annual effective dose for ^{232}Th was found to be three times greater than the UNSCEAR acceptable limits, the total values of annual effective dose were within the acceptable limits, indicating no threat to the consumers. Maximum and minimum values of the total internal hazard index were 0.79 ± 0.06 and 0.04 ± 0.01, respectively, and all the values were less than 1, showing that there is no health hazard to the consumers.

Table 5. Annual effective dose, total effective dose and total internal hazard index for different radionuclides in marine fish and crustacean samples collected at the Bay of Bengal (Bangladesh) during rainy and autumn seasons of 2017.

Sl. No.	Types of Samples	Name of the Organism/Local Name	Scientific Name	No. of Samples	Annual Effective Dose for ^{226}Ra (Svy^{-1})	Annual Effective Dose for ^{232}Th (Svy^{-1})	Annual Effective Dose for ^{40}K (Svy^{-1})	Total Annual Effective Dose (Svy^{-1})	Total Internal Hazard Index (H_{in})
1	Fish	Hilsa fish/ Ilish fish	*Tenualosa ilisha*	2	BDL	3.39×10^{-5}	4.33×10^{-5}	7.72×10^{-5}	0.09 ± 0.01
2		Bombay duck/ Lotia fish	*Harpodon neherues*	2	2.88×10^{-5}	8.96×10^{-4}	3.82×10^{-5}	9.63×10^{-4}	0.74 ± 0.13
3		Flathead sillago/ Sundara Baila	*Sillaginopsis panijus*	2	BDL	9.90×10^{-4}	4.18×10^{-5}	1.03×10^{-3}	0.79 ± 0.06
4		Indian oil sardine/ Colombo fish	*Sardinella longiceps*	2	BDL	BDL	2.97×10^{-5}	2.97×10^{-5}	0.04 ± 0.01
5		Belt fish/ Churi Fish	*Trichiurus lepturus*	2	BDL	5.52×10^{-5}	5.03×10^{-5}	1.05×10^{-4}	0.11 ± 0.02
6		Dotted gizzard shad/ Shard fish	*Konosirus punctatus*	2	BDL	BDL	3.47×10^{-5}	3.47×10^{-5}	0.05 ± 0.02
7	Crustacean	Three-spot swimming crab	*Portunus sanguinolentus*	2	BDL	BDL	2.62×10^{-5}	2.62×10^{-5}	0.04 ± 0.01
8		Mud crab	*Scylla serrata*	2	BDL	2.76×10^{-4}	3.40×10^{-5}	3.10×10^{-4}	0.25 ± 0.05
9		Asian shore crab/ Chati Kakra	*Hemigrapsus takanoi*	2	BDL	1.92×10^{-4}	3.04×10^{-5}	2.22×10^{-4}	0.19 ± 0.03
10		Cat tiger shrimp/ Harina Chingri	*Penaeus semisulcatus*	2	BDL	2.63×10^{-5}	1.90×10^{-5}	4.52×10^{-5}	0.05 ± 0.02

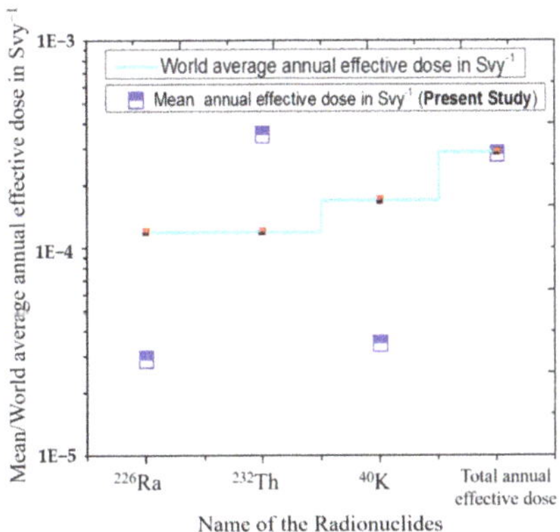

Figure 2. Comparison between the estimated annual effective doses obtained in the present study with the world average values [32].

3.2. Pollution Level and Health Risk Assessment of Trace Metal Analysis

3.2.1. Trace Metal Concentration

The intake, bioassimilation, and subsequent bioaccumulation of trace metals in marine organisms are significantly affected by the dissimilar aquatic geochemistry of the trace metals as well as the diverse feeding habits and living methods of fish and crustaceans. Among marine animals, fish are possibly one of the most mobile and capable of travelling long distances. Compared to other types of organisms, fish are also on a high trophic level in the food chain; hence, their diet is probably the most diverse of the species studied here. Crustaceans are benthic organisms that usually live on or near the sea floor and are capable of travelling some distance. They often feed on organic debris but also on small crustaceans and fish on or near the sea floor. Although the trace metal concentrations in the various species of marine fish and crustaceans analyzed in the present study were in a wide range of variations, these aquatic organisms also showed metal accumulation patterns that were noteworthy (see Table 6).

Table 6. Mean metal concentration (mgkg^{-1} DW) in different marine fish and crustacean samples collected at the Bay of Bengal (Bangladesh) during rainy and autumn seasons of 2017.

Name of the Organism/Local Name	Hilsa Fish/ Ilish Fish	Bombay Duck/ Lotia Fish	Flathead Sillago/ Sundara Baila	Indian Oil Sardine/ Colombo Fish	Belt Fish/ Churi Fish	Dotted Gizzard Shad/ Shard Fish	Three-Spot Swimming Crab	Mud Crab	Asian Shore Crab/ Chati Kakra	Cat Tiger Shrimp/ Harina Chingri
Scientific name	*Tenualosa ilisha*	*Harpodon neherues*	*Sillaginopsis panijus*	*Sardinella longiceps*	*Trichiurus lepturus*	*Konosirus punctatus*	*Portunus sanguinolentus*	*Scylla serrata*	*Hemigrapsus takanoi*	*Penaeus semisulcatus*
Sample Type	Fish						Crustacean			
No. of samples from two seasons	2	2	2	2	2	2	2	2	2	2
Mean Pb conc. in mgkg^{-1} DW	0.19 ± 0.04	0.25 ± 0.03	0.37 ± 0.03	4.6 ± 0.1	0.37 ± 0.00	0.12 ± 0.03	1.4 ± 0.0	1.6 ± 0.0	1.4 ± 0.0	0.12 ± 0.02
Mean Cu conc. in mgkg^{-1} DW	4.1 ± 0.1	1.7 ± 0.1	1.6 ± 0.0	2.6 ± 0.0	1.9 ± 0.0	3.4 ± 0.1	62 ± 20	100 ± 0	75 ± 6	44 ± 0
Mean Zn conc. in mgkg^{-1} DW	73 ± 0	58 ± 0	61 ± 0	170 ± 0	63 ± 0	74 ± 0	130 ± 1	140 ± 1	120 ± 1	83 ± 1
Mean Mn conc. in mgkg^{-1} DW	16 ± 0	5.8 ± 0.0	18 ± 0	20 ± 0	11 ± 0	10 ± 0	89 ± 1	610 ± 1	450 ± 3	17 ± 0
Mean Cd conc. in mgkg^{-1} DW	0.14 ± 0.00	0.16 ± 0.00	0.12 ± 0.00	0.39 ± 0.00	0.11 ± 0.00	2.0 ± 0.0	3.3 ± 0.0	0.52 ± 0.04	0.10 ± 0.00	1.8 ± 0.0
Mean Ni conc. in mgkg^{-1} DW	6.9 ± 0.1	BDL	0.14 ± 0.07	0.37 ± 0.04	1.1 ± 0.1	2.1 ± 0.0	1.6 ± 0.0	1.3 ± 0.1	BDL	0.16 ± 0.02
Mean Cr conc. in mgkg^{-1} DW	0.45 ± 0.00	1.2 ± 0.2	1.8 ± 0.1	0.58 ± 0.00	0.79 ± 0.00	0.79 ± 0.00	2.0 ± 0.2	2.8 ± 0.0	1.8 ± 0.1	0.70 ± 0.00
Mean Fe conc. in mgkg^{-1} DW	790 ± 10	5.5 ± 0.1	140 ± 1	370 ± 2	120 ± 1	130 ± 1	330 ± 1	400 ± 2	560 ± 3	55 ± 0

Note: BDL—below detection limit.

The trace metal concentrations found in individual fish muscles for both rainy and autumn seasons varied for Pb, 7.5–0.12, Cu: 4.3–1.3; Zn, 170–45; Mn, 24–5.5; Cd, 2.1–0.10; Ni, 13–0.28; Cr, 2.6–0.16; and Fe, 1300–11 mgkg^{-1} dry weight. In addition, in individual crustacean samples of both rainy and autumn seasons, these values varied for Pb, 1.9–0.25; Cu, 100–39; Zn, 150–79; Mn, 640–15; Cd, 3.4–0.09; Ni, 2.3–0.31; Cr, 3.3–0.63; and Fe, 600–9.4 mgkg^{-1} dry weight.

Most of the chemical elements present in fish and crustaceans are essential for biota at low concentrations; however, high concentrations of these elements can become toxic for them. Similarly, some metals (e.g., Fe, Mn, Ni, Cu, and Zn) are necessary for proper metabolic reactions in the human body, and other elements (e.g., Cd, Cr, and Pb) are of unknown benefits and may become toxic in the cases of chronic exposure to humans [50]. The obtained trace metal concentrations in the ten marine fish and crustacean samples analyzed in the present study in the rainy season followed the order of Fe > Mn > Zn >

Cu > Ni > Cr > Cd > Pb and in the autumn season followed the order of Fe > Mn > Zn > Cu >Cr > Pb > Cd > Ni. However, in both seasons, the average concentration followed the order of Fe > Zn > Mn > Cu > Ni > Pb > Cr > Cd in fish samples and of Fe > Mn > Zn > Cu > Cr > Cd > Pb > Ni in crustacean samples. This demonstrated that marine organisms in different groups had different accumulation mechanisms for trace metals. Additionally, the different concentrations of chemical elements in the marine environment and the age of the marine organisms analyzed may also be responsible for these trends.

The maximum allowed concentration of Pb in fish is 0.21 $mgkg^{-1}$ DW set by JECFA [51], whereas in our study we found that the fish *Sillaginopsis panijus*, *Sardinella longiceps*, *Trichiurus lepturus*, and all the crustacean species contained Pb concentrations exceeding the limit. According to FAO/WHO, the acceptable limit of Cd for human consumption is 0.20 $mgkg^{-1}$ [52]. Among the fish species, *Sardinella longiceps*, *Konosirus punctatus*, and among the crustacean species *Portunus sanguinolentus*, *Scylla serrata*, and *Penaeus semisulcatus* exceeded this limit. In addition, the permissible limit for Cr is 0.20 $mgkg^{-1}$ DW set by the National Academic Press, Washington DC, 1989 [51], whereas all the studied samples exceeded this limit. These three trace metals (Cd, Cr, and Pb) are considered to be toxic and all of them were in an excessive content in some of the studied marine organisms, demonstrating that the aquatic fauna was contaminated by these metals. Other essential trace elements (e.g., Cu, Zn, Mn, Ni, and Fe) were also observed in excessive contents in some of the studied samples. This indicates the overall pollution status of the studied area due to industrial activities, fossil fuels, agricultural run-off, ship-breaking activities, and other human activities around the coastline.

A comparison between available fish species metal concentration data and the present study is shown in Table 7. It is evident that the range of concentration of metal elements in other parts of the world are lower than that of the presently studied data [52–56]. On the other hand, the obtained data is comparable to similar studies in the neighboring regions [57,58]. If we liken the present study with similar kinds of studies conducted in the Bay of Bengal in the past, the range of metal element concentrations are also comparable [59–61], though it indicates a gradual increment in metal accumulation by the fish of those regions. It seems that day by day, the rate of environmental disturbances and the pollution levels increase due to different reasons, such as the establishment of different industries beside the rivers, which is gradually increasing and also affects the marine environment progressively, contaminating the marine organisms in the long run. Presently, it is an issue of concern for the local populations, but in the near future it might be the threat for the surrounding environment of all organisms, including humans.

Table 7. Trace metal concentrations ($mgkg^{-1}$ DW) in various species of fish from different areas of the world.

Sl. No.	Area	Species	Pb	Cd	Ni	Cr	Cu	Zn	Mn	Fe	Ref.
1	Bay of Bengal, Bangladesh	Five species: *P. carcinus*, *W. attu*, *S. sinensis*, *R. rita*, *S. sihama*, and *B. strogylurus*	-	-	2.70–15.20	-	0.65–66.00	26.00–78.80	1.89–7.11	37.90–182.00	[61]
2	Bay of Bengal, Bangladesh	Six species: *C. neglecta*, *C. reba*, *J. argentus*, *H. nehereus*, *S. fhasa*, and *L. savala*	1.67–2.58	0.009–0.17	6.44–7.58	-	3.33–4.69	18.86–33.89	5.01–11.14	60.55–451.10	[59]
3	Bay of Bengal, Bangladesh	Nine species: *L. calcarifer*, *P. pangasius*, *P. indicus*, *I. megaloptera*, *R. russelliana*, *L. stylifetus*, *R. kanagurta*, *S. guttatum*, *P. paradiseus*, and *O. militaris*	0.58–4.03	0.04–0.13	0.73–6.11	-	0.46–7.74	25.67–119.36	1.46–9.02	37.70–118.91	[60]

Table 7. *Cont.*

Sl. No.	Area	Species	Pb	Cd	Ni	Cr	Cu	Zn	Mn	Fe	Ref.
4	Northeast Coast of the Bay of Bengal, India	Five species: *S. phasa*, *P. argentus*, *G. sparsipapillus*, *L. parsia*, and *C.* sp.	12.40–19.96	0.62–1.20	2.20–3.69	ND–3.89	16.22–47.97	12.13–44.74	-	-	[57]
5	Cuddalore Coast, Tamil Nadu, India	Three species: *R. kanagurta*, *K. axillaris*, and *S. longiceps*	-	0.35–0.43	0.62–0.79	0.66–0.86	0.42–0.61	20.10–26.20	-	-	[56]
6	Rio de Janeiro, Brazil	Eleven species: *S. salar*, *S. brasiliensis*, *P. saltatrix*, *M. furnieri*, *C. leiarchus*, *C. crysos*, *P. arenatus*, *M. cephalus*, *G. brasiliensis*, *L. villarii*, and *P. numida*	0.04–0.30	0.001–0.09	-	-	1.20–2.90	2.70–9.30	0.30–1.70	1.60–7.50	[55]
7	Northeast Coast of India	Nine species: *H. nehereus*, *T. trichiurus*, *S. laticaudus*, *D. albida*, *P. argentius*, *A.* sp., *F. niger*, *H. ilisha*, and *R. kanagurta*	-	0.01–1.10	-	-	0.50–28.20	3.00–99.10	0.50–12.00	10.40–249.70	[58]
8	Mersing, Malaysia	Two species: *A. thalassinus*, and *J. belangeri*	-	2.20–2.34	-	-	8.80–12.91	120.91–217.37	4.34–9.67	-	[52]
9	Saudi Arabia	Three species: *L. nebulosus*, *P. major*, and *S. cantharu*	0.002–0.003	0.001–0.001	-	-	0.026–0.093	0.037–0.376	0.008–0.036	0.222–1.016	[54]
10	Bay of Bengal, Bangladesh	Ten species: *L. calcarifer*, *P. pangasius*, *P. indicus*, *I. megaloptera*, *A. cruciger*, *P. chinensis*, *S. phasa*, *S. guttatum*, *C. reba*, and *A. arius*	0.80–6.26	0.02–0.47	1.88–7.56	1.27–4.66	≤ 8.54	13.22–74.36	3.63–17.80	-	[35]
11	Terengganu Coastal Area, Malaysia	Ten species: *S. leptolepis*, *D. maraudsi*, *E. lanceolatus*, *P. tayenus*, *Rastrelliger*, *M. cordyla*, *N. soldado*, *P. filamentosus*, *Bramidae*, and *S. canaliculatus*	0.0002–0.007	0.0008–0.015	-	-	0.0021–0.012	0.0488–0.151	0.0068–0.041	0.3995–0.667	[53]
12	Northwest Mediterranean Sea	Six species: *G. melastomus*, *S. canicula*, *H. dactylopterus*, *L. boscii*, *M. poutassou*, and *P. blennoides*	0.00–0.90	0.00–0.03	0.02–7.00	-	0.10–10.60	12.10–60.30	-	-	[7]
13	Northeast Atlantic Ocean	Six species: *G. melastomus*, *S. canicula*, *H. dactylopterus*, *L. boscii*, *M. poutassou*, and *P. blennoides*	0.01–0.26	0.00–0.04	0.00–0.53	-	0.40–5.80	9.90–40.00	-	-	[7]
14	Bay of Bengal, Bangladesh	Six species: *T. ilisha*, *H. neherues*, *S. panijus*, *S. longiceps*, *T. lepturus*, and *K. punctatus*	0.12–4.6	0.11–2	BDL–6.9	0.45–1.8	1.6–4.1	58–170	5.8–20	5.5–790	PS

Note: BDL—below detection limit; ND—not detected; PS—present study.

3.2.2. Estimated Daily Intake (EDI)

EDI and mean EDI values of respective trace elements (Cd, Pb, Zn, Cu, Ni, Fe, Mn, and Cr) for the fish and crustacean samples of both seasons are shown in Table 8. The corresponding Figure 3 shows the comparative results of estimated daily intake (EDI) values with the recommended daily intake (RDI) values. As the values of EDI < RDI for both fish and crustaceans, there is no risk from the consumption of the studied seafood.

Table 8. The values of estimated daily intake (EDI), target hazard quotients (THQ), and hazard index (HI) obtained for the marine fish and crustacean samples collected at the Bay of Bengal (Bangladesh) during rainy and autumn seasons of 2017.

Sampling Time	Rainy Season			Autumn Season			Mean EDI ($mgkg^{-1}$ Body Weight per Day)	Mean THQ	HI
Trace Metals	Average Concentration ($mgkg^{-1}$) DW	EDI ($mgkg^{-1}$ Body Weight per Day)	THQ	Average Concentration ($mgkg^{-1}$) DW	EDI ($mgkg^{-1}$ Body Weight per Day)	THQ			
Pb	0.79	8.29×10^{-4}	4.31×10^{-5}	1.26	1.31×10^{-3}	8.54×10^{-5}	1.07×10^{-3}	6.43×10^{-5}	8.82×10^{-4}
Cu	27.20	2.84×10^{-2}	1.48×10^{-4}	33.10	3.45×10^{-2}	1.80×10^{-4}	3.15×10^{-2}	1.64×10^{-4}	
Zn	93.92	9.80×10^{-2}	6.79×10^{-5}	98.69	1.03×10^{-1}	7.14×10^{-5}	1.01×10^{-1}	6.96×10^{-5}	
Mn	122.51	1.28×10^{-1}	1.90×10^{-4}	128.20	1.34×10^{-1}	1.99×10^{-4}	1.31×10^{-1}	1.94×10^{-4}	
Cd	0.84	8.80×10^{-4}	1.83×10^{-4}	0.88	9.18×10^{-4}	1.91×10^{-4}	8.99×10^{-4}	1.87×10^{-4}	
Ni	1.93	2.01×10^{-3}	2.09×10^{-5}	0.78	8.09×10^{-4}	1.20×10^{-5}	1.41×10^{-3}	1.65×10^{-5}	
Cr	1.29	1.35×10^{-3}	9.36×10^{-5}	1.26	1.32×10^{-3}	9.15×10^{-5}	1.34×10^{-3}	9.26×10^{-5}	
Fe	308.62	3.22×10^{-1}	9.56×10^{-5}	272.24	2.84×10^{-1}	9.37×10^{-5}	3.03×10^{-1}	9.47×10^{-5}	

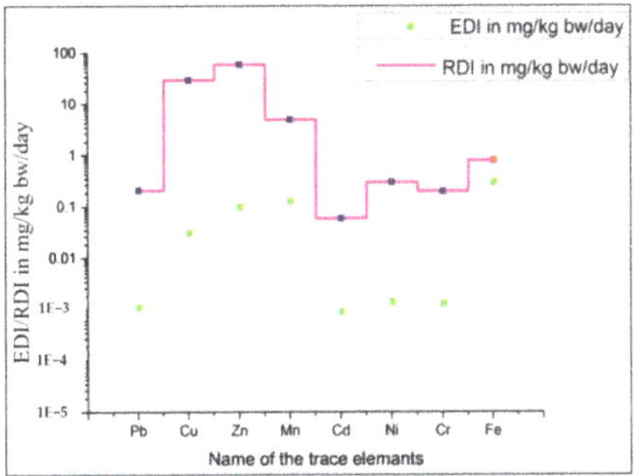

Figure 3. Comparison between estimated daily intake (EDI) values and recommended daily intake (RDI) values obtained in the present study.

3.2.3. Target Hazard Quotients (THQ)

The dimensionless target hazard quotient (THQ) is the indicator of non-carcinogenic health risks associated with the consumption of food (fish and crustaceans). The mean THQ values for the trace metals Pb, Cu, Zn, Mn, Cd, Ni, Cr and Fe are given in Table 8. After analysis of all the fish and crustacean samples, the obtained THQ values for different metals were less than 1, which indicates that the exposed population is unlikely to experience obvious adverse effects [39].

3.2.4. Hazard Index (HI)

The value of the hazard index, which is obtained by the summing of the target hazard quotients of each metal, was used to assess the overall potential health risk posed by more than one metal. The obtained hazard index value was 8.82×10^{-4}, which is less than 1, indicating that there are no health hazards to the local consumers (Table 8).

3.2.5. Target Cancer Risk (TR)

The obtained values of target cancer risk for Pb, Cd, Ni, and Cr due to exposure from the consumption of the targeted six fish and four crustacean species analyzed in the present study are shown in Figure 4. Generally, the values of TR lower than 10^{-6} are considered as negligible, while those above 10^{-4} are considered to be unacceptable, and those in between 10^{-6} and 10^{-4} are considered as an acceptable range [37]. The present study reveals that only TR for Pb were below the benchmark and those for Cd, Ni, and Cr were above the benchmark, indicating that the fish and the crustaceans were becoming polluted. This also increased the risk of chronic cancer due to exposure of Cd, Ni, and Cr through fish and crustacean consumption.

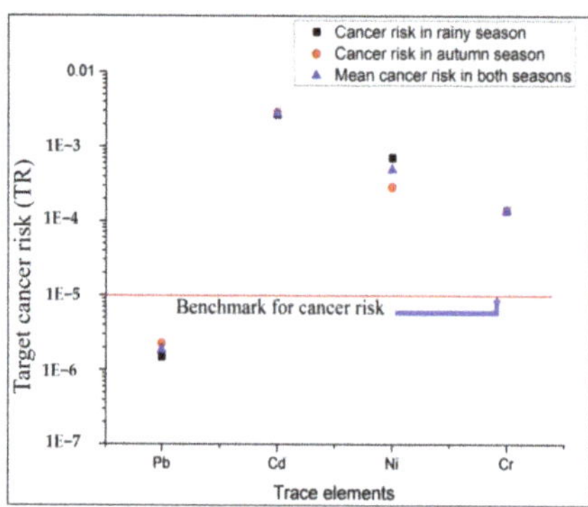

Figure 4. Targeted cancer risk (TR) values for associated toxic element consumption from the studied seafood (fish and crustaceans) at the Bay of Bengal (Bangladesh) during rainy and autumn seasons of 2017.

4. Conclusions

The radioactivity and the trace metal concentrations were studied in certain commercially important marine biota (fish and crustaceans) from the Bay of Bengal, Bangladesh, in two different seasons (rainy and autumn). Through the analysis of different radiological parameters, the present study revealed an elevated activity concentration compared to the acceptable limits in some of the samples. Comparing the present results with the reported results of the different regions of the world, including similar studies at the Bay of Bengal, it can be concluded that the studied region possesses more radioactive elements than the other marine regions. It was also observed that the marine radioactivity of the studied region is gradually increasing, day by day. The results reflect the contribution of technologically enhanced naturally occurring radioactive material (TENORM) pollutants, largely expected to be a result of power generation plants, petroleum, steel, shipbuilding, chemical, pharmaceutical, textile, vegetable oil refineries, glass manufacturing industries, etc. However, the present study indicates that radionuclide intake from the consumption of Bay of Bengal fish and crustaceans still poses an insignificant threat to public health.

Conversely, the metal concentrations of a few trace elements are higher than the acceptable limits in most of the samples, although the studied fish and crustaceans are still safe for human consumption. However, the target cancer risk (TR> 10^{-4}) due to exposure to cadmium, nickel, and chromium indicates that consumer risk persists.

As a whole, the present study indicates an increase in the pollution by radioactivity and trace metals in the marine environment of the Bay of Bengal, mostly derived from an increase in human activities in the region. Therefore, consuming the seafood from the studied region has the potential to cause adverse health impacts if not controlled, and at the same time, all the stakeholders should take proper initiatives to prevent environmental pollution of the Bay of Bengal. The results reported here can be used as baseline data, though the sample size is not large enough. Similar studies should be planned and carried out periodically with higher sample sizes to monitor any changes in marine pollution of the Bay of Bengal in the future. Statistical analyses should also be employed in future studies for better understanding of the scenario.

Author Contributions: Conceptualization, K.P.B., A.K.M.S.I.B., S.H.2 and M.B.H.; methodology, S.H.1, N.D., A.K.M.S.I.B. and S.H.2; software, K.P.B., S.H.1 and N.D.; validation, A.K.M.S.I.B., S.C.G., S.H.2 and M.B.H.; formal analysis, K.P.B., S.H.1 and N.D.; investigation, K.P.B., S.H.1, N.D. and A.K.M.S.I.B.; resources, A.K.M.S.I.B. and S.H.2; data curation, K.P.B., S.H.1 and N.D.; writing—original draft preparation, K.P.B. and A.K.M.S.I.B.; writing—review and editing, S.H.1, A.K.M.S.I.B. and S.C.G.; visualization, A.K.M.S.I.B. and S.H.2; supervision, A.K.M.S.I.B., S.H.2 and M.B.H.; and project administration, S.H.2 and M.B.H. All authors have read and agreed to the published version of the manuscript.

Funding: This research received no external funding.

Data Availability Statement: The data presented in this study are contained within the article.

Acknowledgments: The authors are thankful to the authority of the Atomic Energy Centre, Chattogram, Bangladesh Atomic Energy Commission, for providing support and technical co-operation to analyze the specimens using the HPGe and AAS. The authors are also delighted to express their gratefulness to the Ministry of Shipping, Peoples' Republic of Bangladesh, for their utmost support. Thanks also goes to Masuda Begum Sampa, Kyushu Institute of Technology, Japan for the fruitful discussion.

Conflicts of Interest: The authors declare no conflict of interest.

References

1. Ghorab, M.A. Environmental pollution by heavy metals in the aquatic ecosystems of Egypt. *Open Access J. Toxicol.* **2018**, *3*, 555603. [CrossRef]
2. Bhuiyan, M.; Ali, M.; Rahman, M. Marine boundary confirmation of Bangladesh: Potentials of sea resources and challenges ahead. *Cost Manag.* **2015**, *43*, 18–24.
3. Alam, M.K.; Chakraborty, S.R.; Rahman, A.R.; Deb, A.K.; Kamal, M.; Bhuian, A.S.I. Measurement of radioactivity and hence define the radiological risk associated with the Chittagong city site coastal sediment containing all types of wastes (mills, factories, industries, and municipalities) in Bangladesh. *Radiat. Prot. Environ.* **2011**, *34*, 4–8.
4. Rashid, T.; Hoque, S.; Akter, S. Pollution in the bay of Bengal: Impact on marine ecosystem. *Open J. Mar. Sci.* **2015**, *5*, 55–63. [CrossRef]
5. Solomon, K. Sources of radioactivity in the ocean environment: From low level waste to nuclear powered submarines. *J. Hazard. Mater.* **1988**, *18*, 255–262. [CrossRef]
6. Fazio, F.; Saoca, C.; Ferrantelli, V.; Cammilleri, G.; Capillo, G.; Piccione, G. Relationship between arsenic accumulation in tissues and hematological parameters in mullet caught in Faro Lake: A preliminary study. *Environ. Sci. Pollut. Res.* **2019**, *26*, 8821–8827. [CrossRef]
7. Mille, T.; Cresson, P.; Chouvelon, T.; Bustamante, P.; Brach-Papa, C.; Bruzac, S.; Rozuel, E.; Bouchoucha, M.; Tiphaine, M.; Pierre, C.; et al. Trace metal concentrations in the muscle of seven marine species: Comparison between the Gulf of Lions (North-West Mediterranean Sea) and the Bay of Biscay (North-East Atlantic Ocean). *Mar. Pollut. Bull.* **2018**, *135*, 9–16. [CrossRef]
8. Fazio, F.; Piccione, G.; Tribulato, K.; Ferrantelli, V.; Giangrosso, G.; Arfuso, F.; Faggio, C. Bioaccumulation of heavy metals in blood and tissue of striped mullet in two italian lakes. *J. Aquat. Anim. Health* **2014**, *26*, 278–284. [CrossRef]
9. Cammilleri, G.; Vazzana, M.; Arizza, V.; Giunta, F.; Vella, A.; Dico, G.L.; Giaccone, V.; Giofrè, S.V.; Giangrosso, G.; Cicero, N.; et al. Mercury in fish products: What's the best for consumers between bluefin tuna and yellowfin tuna? *Nat. Prod. Res.* **2018**, *32*, 457–462. [CrossRef]
10. Cammilleri, G.; Galluzzo, F.G.; Fazio, F.; Pulvirenti, A.; Vella, A.; Dico, G.M.L.; Macaluso, A.; Ciaccio, G.; Ferrantelli, V. Mercury detection in benthic and pelagic fish collected from Western Sicily (Southern Italy). *Animals* **2019**, *9*, 594. [CrossRef]
11. Anandkumar, A.; Nagarajan, R.; Prabakaran, K.; Bing, C.H.; Rajaram, R.; Li, J.; Du, D. Bioaccumulation of trace metals in the coastal Borneo (Malaysia) and health risk assessment. *Mar. Pollut. Bull.* **2019**, *145*, 56–66. [CrossRef]
12. Khandaker, M.U.; Asaduzzaman, K.; Nawi, S.M.; Usman, A.R.; Amin, Y.M.; Daar, E.; Bradley, D.A.; Ahmed, H.; Okhunov, A.A. assessment of radiation and heavy metals risk due to the dietary intake of marine fishes (*Rastrelliger kanagurta*) from the straits of Malacca. *PLoS ONE* **2015**, *10*, e0128790. [CrossRef]
13. Kalay, M.; Canli, M. Elimination of essential (Cu, Zn) and non-essential (Cd, Pb) metals from tissues of a freshwater fish Tilapia zilli. *Turk. J. Zool.* **2000**, *24*, 429–436.
14. Gorur, F.K.; Keser, R.; Akçay, N.; Dizman, S. Radioactivity and heavy metal concentrations of some commercial fish species consumed in the Black Sea Region of Turkey. *Chemosphere* **2012**, *87*, 356–361. [CrossRef] [PubMed]

15. Zalewska, T.; Suplińska, M. Fish pollution with anthropogenic 137Cs in the southern Baltic Sea. *Chemosphere* **2013**, *90*, 1760–1766. [CrossRef] [PubMed]
16. Cantillo, A. Comparison of results of Mussel Watch Programs of the United States and France with Worldwide Mussel Watch Studies. *Mar. Pollut. Bull.* **1998**, *36*, 712–717. [CrossRef]
17. i Batlle, J.V. Radioactivity in the Marine Environment. In *Encyclopedia of Sustainability Science and Technology*; Meyers, R.A., Ed.; Springer: New York, NY, USA, 2012; pp. 8387–8425.
18. Elnabris, K.J.; Muzyed, S.K.; El-Ashgar, N.M. Heavy metal concentrations in some commercially important fishes and their contribution to heavy metals exposure in Palestinian people of Gaza Strip (Palestine). *J. Assoc. Arab. Univ. Basic Appl. Sci.* **2013**, *13*, 44–51. [CrossRef]
19. Khandaker, M.U.; Heffny, N.; Adillah, B.; Amin, Y.M.; Bradley, D. Elevated concentration of radioactive potassium in edible algae cultivated in Malaysian seas and estimation of ingestion dose to humans. *Algal Res.* **2019**, *38*, 101386. [CrossRef]
20. Khandaker, M.U.; Uwatse, O.B.; Khairi, K.A.B.S.; Faruque, M.R.I.; Bradley, D.A. Terrestrial radionuclides in surface (DAM) water and concomitant dose in metropolitan Kuala Lumpur. *Radiat. Prot. Dosim.* **2019**, *185*, 343–350. [CrossRef]
21. Doulah, N.-U.; Karim, M.R.; Hossain, S.; Deb, N.; Barua, B.S. Spatial distribution of heavy metals in surface and sub surface sediments of the coastal area of Kutubdia Island, Cox's Bazar, Bangladesh. *J. Environ. Anal. Chem.* **2017**, 4. [CrossRef]
22. Rahman, M.S.; Barua, B.S.; Karim, R.; Kamal, M. Investigation of heavy metals and radionuclide's impact on environment due to the waste products of different iron processing industries in Chittagong, Bangladesh. *J. Environ. Prot.* **2017**, *8*, 974–989. [CrossRef]
23. Currie, L.A. Nomenclature in evaluation of analytical methods including detection and quantification capabilities (IUPAC Recommendations 1995). *Pure Appl. Chem.* **1995**, *67*, 1699–1724. [CrossRef]
24. Eaton, A.D.; Clesceri, L.S.; Greenberg, A.E.; Franson, M.A.H. *Standard Methods for the Examination of Water and Wastewater*, 22nd ed.; American Public Health Association APHA: Washington, DC, USA, 2012.
25. Amin, R.M.; Ahmed, F. Estimation of annual effective dose to the adult Egyptian population due to natural radioactive elements in ingestion of spices. *Adv. Appl. Sci. Res.* **2013**, *4*, 350–354.
26. ICRP. *Dose Coefficents for Intakes of Radionuclides by Workers*; Publication 68; ICRP: Ottawa, ON, Canada, 1994.
27. Pulhani, V.; Dafauti, S.; Hegde, A.; Sharma, R.; Mishra, U. Uptake and distribution of natural radioactivity in wheat plants from soil. *J. Environ. Radioact.* **2005**, *79*, 331–346. [CrossRef]
28. DoF. *Yearbook of Fisheries Statistics of Bangladesh 2017–18*; Fisheries Resources Survey System (FRSS), Department of Fisheries, Ministry of Fisheries: Dhaka, Bangladesh , December 2018; Volume 35, p. 86.
29. Aliyu, A.O.; Yakubu, A.; Ajagbe, O.O. Determination of radionuclide concentrations, hazard indices and physiochemical parameters of water, fishes and sediments in River Kaduna, Nigeria. *IOSR J. Appl. Chem.* **2018**, *11*, 28–34. [CrossRef]
30. Diab, H.M.; Nouh, S.A.; Hamdy, A.; EL-Fiki, S.A. Evaluation of natural radioactivity in a cultivated area around a fertilizer factory. *J. Nucl. Radiat. Phys.* **2008**, *3*, 53–62.
31. Pawel, D.; Leggett, R.W.; Eckerman, K.F.; Nelson, C.B. *Uncertainties in Cancer Risk Coefficients for Environmental Exposure to Radionuclides. An Uncertainty Analysis for Risk Coefficients Reported in Federal Guidance Report No. 13*; Oak Ridge National Laboratory: Oak Ridge, TN, USA, 2007.
32. Asaduzzaman, K.; Khandaker, M.; Amin, Y.; Mahat, R. Uptake and distribution of natural radioactivity in rice from soil in north and west part of peninsular Malaysia for the estimation of ingestion dose to man. *Ann. Nucl. Energy* **2015**, *76*, 85–93. [CrossRef]
33. BBS. *Statistical Year Book Bangladesh 2018*; BBS: Dhaka, Bangladesh, 2019.
34. Patra, A.C.; Mohapatra, S.; Sahoo, S.K.; Lenka, P.; Dubey, J.S.; Tripathi, R.M.; Puranik, V.D. Age-dependent dose and health risk due to intake of uranium in drinking water from Jaduguda, India. *Radiat. Prot. Dosim.* **2013**, *155*, 210–216. [CrossRef]
35. Saha, N.; Mollah, M.; Alam, M.; Rahman, M.S. Seasonal investigation of heavy metals in marine fishes captured from the Bay of Bengal and the implications for human health risk assessment. *Food Control* **2016**, *70*, 110–118. [CrossRef]
36. Akoto, O.; Bismark Eshun, F.; Darko, G.; Adei, E. Concentrations and health risk assessments of heavy metals in fish from the Fosu Lagoon. *Int. J. Environ. Res.* **2014**, *8*, 403–410.
37. Ahmed, K.; Shaheen, N.; Islam, S.; Mamun, H.; Mohiduzzaman, A.; Bhattacharjee, L. Dietary intake of trace elements from highly consumed cultured fish (Labeo rohita, Pangasius pangasius and Oreochromis mossambicus) and human health risk implications in Bangladesh. *Chemosphere* **2015**, *128*, 284–292. [CrossRef]
38. USEPA. In *Risk Assessment Guidance for Superfund—Human Health Evaluation Manual (Part A)*; U.S. Environmental Protection Agency: Washington, DC, USA, 1989; 1, pp. 1–288.
39. Wang, X.; Sato, T.; Xing, B.; Tao, S. Health risks of heavy metals to the general public in Tianjin, China via consumption of vegetables and fish. *Sci. Total Environ.* **2005**, *350*, 28–37. [CrossRef] [PubMed]
40. Javed, M.; Usmani, N. Accumulation of heavy metals and human health risk assessment via the consumption of freshwater fish Mastacembelus armatus inhabiting, thermal power plant effluent loaded canal. *SpringerPlus* **2016**, *5*, 776. [CrossRef]
41. Zeng, F.; Wei, W.; Li, M.; Huang, R.; Yang, F.; Duan, Y. Heavy metal contamination in rice-producing soils of Hunan Province, China and potential health risks. *Int. J. Environ. Res. Public Health* **2015**, *12*, 15584–15593. [CrossRef]
42. Pollution of Karnafully River Water—Causes, Effects and Recommendations. 2003. Available online: https://gspmarinebd.com/wp-content/uploads/2020/01/River.pdf (accessed on 17 December 2020).

43. Yasmin, S.; Barua, B.S.; Khandaker, M.U.; Kamal, M.; Rashid, A.; Sani, S.A.; Ahmed, H.; Nikouravan, B.; Bradley, D. The presence of radioactive materials in soil, sand and sediment samples of Potenga sea beach area, Chittagong, Bangladesh: Geological characteristics and environmental implication. *Results Phys.* **2018**, *8*, 1268–1274. [CrossRef]
44. IAEA. *Sediment Distribution Coefficients and Concentration Factors for Biota in the Marine Environment*; Technical Report No. 422; IAEA: Wien, Austria, 2004.
45. Khan, M.F.; Wesley, S.G. Radionuclides in resident and migratory fishes of a wedge bank region: Estimation of dose to human beings, South India. *Mar. Pollut. Bull.* **2012**, *64*, 2224–2232. [CrossRef]
46. Sowol, O.; Giwa, K.W. Estimation of annual dose rate of natural radionuclides in man from crabs in River Odeomi Ijebu waterside Ogun State Nigeria. *Int. J. Sci. Technol.* **2013**, *2*, 438–441.
47. Alam, M.; Chowdhury, M.; Kamal, M.; Ghose, S. Radioactivity in marine fish of the Bay of Bengal. *Appl. Radiat. Isot.* **1995**, *46*, 363–364. [CrossRef]
48. Ghose, S.; Alam, M.; Islam, M. Radiation dose estimation from the analysis of radionuclides in marine fish of the Bay of Bengal. *Radiat. Prot. Dosim.* **2000**, *87*, 287–291. [CrossRef]
49. Kabir, M.H.; Miah, M.M.H.; Rahman, M.H.; Kamal, M.; Chowdhury, M.T. Study of the naturally occurring radionuclide concentrations and the estimation of dose rates for the samples collected from the St. Martin's Island, Chittagong, Bangladesh. *Materials A* **2017**, *2*, 39–48.
50. Al-Busaidi, M.; Yesudhason, P.; Almughairi, S.; Alrahbi, W.A.K.; Al-Harthy, K.; Al-Mazrooei, N.; Al-Habsi, S. Toxic metals in commercial marine fish in Oman with reference to national and international standards. *Chemosphere* **2011**, *85*, 67–73. [CrossRef] [PubMed]
51. Ullah, A.K.M.A.; Maksud, M.; Khan, S.; Lutfa, L.; Quraishi, S.B. Dietary intake of heavy metals from eight highly consumed species of cultured fish and possible human health risk implications in Bangladesh. *Toxicol. Rep.* **2017**, *4*, 574–579. [CrossRef] [PubMed]
52. Bashir, F.H.; Othman, M.S.; Mazlan, A.G.; Rahim, S.M.; Simon, K.D. Heavy metal concentration in fishes from the coastal waters of Kapar and Mersing, Malaysia. *Turk. J. Fish. Aquat. Sci.* **2013**, *13*, 375–382. [CrossRef]
53. Rosli, M.; Samat, S.; Yasir, M.; Yusof, M. Analysis of heavy metal accumulation in fish at terengganu coastal area, Malaysia. *Sains Malays.* **2018**, *47*, 1277–1283. [CrossRef]
54. El-Bahr, S.M.; Abdelghany, A. Heavy metal and trace element contents in edible muscle of three commercial fish species, and assessment of possible risks associated with their human consumption in Saudi Arabia. *J. Adv. Vet. Anim. Res.* **2015**, *2*, 271. [CrossRef]
55. Medeiros, R.J.; Dos Santos, L.M.G.; Freire, A.S.; Santelli, R.E.; Braga, A.M.C.; Krauss, T.M.; Jacob, S.D.C. Determination of inorganic trace elements in edible marine fish from Rio de Janeiro State, Brazil. *Food Control* **2012**, *23*, 535–541. [CrossRef]
56. Vijayakumar, P.; Lavanya, R.; Veerappan, N.; Balasubramanian, T. Heavy Metal Concentrations in Three Commercial Fish Species in Cuddalore Coast, Tamil Nadu, India. *J. Exp. Sci.* **2011**, *2*, 20–23.
57. De, T.K.; De, M.; Das, S.; Ray, R.; Ghosh, P.B. Level of heavy metals in some edible marine fishes of mangrove dominated tropical estuarine areas of Hooghly river, North East Coast of Bay of Bengal, India. *Bull. Environ. Contam. Toxicol.* **2010**, *85*, 385–390. [CrossRef]
58. Kumar, B.; Sajwan, K.S.; Mukherjee, D.P. Distribution of heavy metals in valuable coastal fishes from North East Coast of India. *Turk. J. Fish. Aquat. Sci.* **2012**, *12*, 81–88. [CrossRef]
59. Sharif, A.; Mustafa, A.; Mirza, A.; Safiullah, S. Trace metals in tropical marine fish from the Bay of Bengal. *Sci. Total Environ.* **1991**, *107*, 135–142. [CrossRef]
60. Sharif, A.; Mustafa, A.; Amin, M.; Safiullah, S. Trace element concentrations in tropical marine fish from the Bay of Bengal. *Sci. Total Environ.* **1993**, *138*, 223–234. [CrossRef]
61. Khan, A.; Ali, M.; Biaswas, S.; Hadi, D. Trace elements in marine fish from the Bay of Bengal. *Sci. Total Environ.* **1987**, *61*, 121–130. [CrossRef]

environments

MDPI

Article

A Global Meta-Analysis for Estimating Local Ecosystem Service Value Functions

Luiz Magalhães Filho [1,2,*], Peter Roebeling [1,3], Maria Isabel Bastos [1], Waldecy Rodrigues [4] and Giulia Ometto [5]

1 CESAM & Department of Environment & Planning (DAO), University of Aveiro (UA), 3810-193 Aveiro, Portugal; peter.roebeling@ua.pt (P.R.); mariaisabel@ua.pt (M.I.B.)
2 Federal Institute of Education, Science and Technology of Tocantins (IFTO), Dianopolis 77300-000, Brazil
3 Wageningen Economic Research, Wageningen University and Research (WUR), 6708 PB Wageningen, The Netherlands
4 Regional Development Program, Federal University of Tocantins (UFT), Palmas 77001-090, Brazil; waldecy@uft.edu.br
5 Department of Economics, Social Studies, Applied Mathematics and Statistics (DE), University of Turin (UNITO), 10124 Turin, Italy; giulia.ometto@outlook.it
* Correspondence: luizlacerda@ua.pt

Abstract: The Meta-analysis has increasingly been used to synthesize the ecosystem services literature, with some testing of the use of such analyses to transfer benefits. These are typically based on local primary studies. However, meta-analyses associated with ecosystem services are a potentially powerful tool for transferring benefits, especially for environmental assets for which no primary studies are available. In this study we use the Ecosystem Service Valuation Database (ESVD), which brings together 1350 value estimates from more than 320 studies around the world, to estimate meta-regression functions for Provisioning, Regulating and maintenance, and Cultural ecosystem services across 12 biomes. We tested the reliability of these meta-regression functions and found that even using variables with high explanatory power, transfer errors could still be large. We show that meta-analytic transfer performs better than simple value transfer and, in addition, that local meta-analytical transfer (i.e., based on local explanatory variable values) provides more reliable estimates than global meta-analytical transfer (i.e., based on mean global explanatory variable values). Thus, we conclude that when taking into account the characteristics of the study area under analysis, including explanatory variables such as income, population density, and protection status, we can determine the value of ecosystem services with greater accuracy.

Keywords: ecosystem services; benefit transfer; meta-analysis; meta-regression function

Citation: Magalhães Filho, L.; Roebeling, P.; Bastos, M.I.; Rodrigues, W.; Ometto, G. A Global Meta-Analysis for Estimating Local Ecosystem Service Value Functions. *Environments* **2021**, *8*, 76. https://doi.org/10.3390/environments8080076

Academic Editor: Sílvia C. Gonçalves

Received: 18 June 2021
Accepted: 4 August 2021
Published: 9 August 2021

1. Introduction

Jean-Baptiste Say poses the idea of nature's services as costless, free gifts of nature as follows: "the wind which turns our mills, and even the heat of the sun, work for us; but happily no one has yet been able to say, the wind and the sun are mine, and the service which they render must be paid for" [1]. However, currently, it is possible to observe that the overuse or misuse of some natural resources poses direct impacts on society. In the face of this problem came the concept of ecosystem services (ES), defined as the benefits that humans obtain from the natural environment and from properly-functioning ecosystems. Hence, several authors [2–8] argue that the sustainable management of natural resources requires correct valuation of the ecosystem defining their services to the society.

Ecosystem services support human life every day and contribute to human well-being in many ways, which are hard to define in a single notion. Hence, the Millennium Ecosystem Assessment [9] and the Common International Classification of Ecosystem Services [10] differentiate between the following ecosystem services: (a) Provisioning services (such as the supply of food via fishery production, fuel, wood, energy resources, and natural products); (b) Regulating and maintenance services (such as shoreline protection, nutrient regulation, carbon sequestration, detoxification of polluted waters, and waste disposal); and (c) Cultural services (such as tourism and recreation).

Ecosystems have great importance across many dimensions (ecological, socio-cultural, and economic) [5]. Thus, expressing the value of ecosystem services in monetary units (i.e., ecosystem service values; ESV), can prove to be of utmost importance to help raise consciousness and convey the (relative) importance of ecosystems and biodiversity to decision-makers. Indeed, monetized valuation pushes for more efficient use of limited resources and helps to select where protection and regeneration are economically more important and can be delivered at least cost [11,12]. It can also assist in determining "a fair compensation" to be paid for a loss of ES in liability regimes [5].

Historically, in the late 1990s and early 2000s, the concept of ES slowly found its way into the policy arena, e.g., through the "Ecosystem Approach" and the Global Biodiversity Assessment. In 2005, the concept of ES gained wider interest after the publication of the Millennium Ecosystem Assessment by the United Nations for policymakers [4,9]. ES are also entering the consciousness of mainstream media and business, namely through the World Business Council for Sustainable Development that has actively supported and developed this concept [13]. Many projects and groups are currently working toward better understanding, modeling, valuing, and managing ES and natural capital [4].

An increasing number of papers seeking the valuation of ES have been published over the last decades. Assessments have been conducted at local [14–16], national [17–19], continental [20,21], and global [4–6] scales. In the same way, databases compiling data from these primary valuation studies were created to aggregate information and facilitate public debate and policy action. Some examples of such databases include the Economic Valuation Reference Inventory [22], and the Ecosystem Service Valuation Database (ESVD; [23]).

Since the early 1990s, several researchers have investigated the applicability and the precision of benefit transfer. However, these past studies were primarily concerned with traditional methods of benefit transfer (in particular value transfer), replacing values directly from the study site to the policy site without amendments [24]. However, in the late 1990s meta-analysis started to be used, with multivariate regression being investigated for use in benefit transfer [25].

The meta-analysis (MA) is a technique that uses statistical models (meta-regressions) to summarize and evaluate previous research results. In benefit transfer, meta-regression results may be used qualitatively, to corroborate new primary results, or to transfer estimated values [26]. Meta-regression in benefit transfer summarizes the weight of the evidence and characterizes the degree of uncertainty about quality-adjusted ecosystem values. In meta-regression, the value estimates from primary valuation studies are thereby treated as individual observations [27]. Meta-regression also extends the range of primary valuation studies by allowing the estimation of values for services and functions that are constant within each primary valuation study but vary across different valuation studies [28].

Meta-analyses have been performed for specific ecosystem services, biomes, and locations. For example, Van Houtven et al. [15] assessed the cultural value of surface water quality in the United States, using 131 willingness-to-pay (WTP) estimates from 18 studies. Similarly, Hjerpe et al. [29] synthesized 127 WTP estimates from 22 different studies that provided estimations for preservation, forest restoration, and freshwater restoration also in the United States. Ghermandi et al. [30] performed a meta-analysis to determine the values of goods and services provided by wetland ecosystems, using 418 value observations derived from 170 valuation studies and 186 wetland sites worldwide. Finally, Hynes et al. [31] performed a marine recreational meta-analysis estimation, using 311 distinct value observations from 96 primary valuation studies. Nevertheless, there are no studies with a broader analysis, that estimate global meta-regression functions for Provisioning, Regulating and maintenance and Cultural ecosystem services across biomes and continents. In addition, testing the reliability of estimated meta-regression functions is relatively rare, (e.g., [32,33]). One of the main challenges is developing equations for ES that capture the local/regional characteristics of the biome and provide reliable value estimates.

Hence, the objective of this study is to estimate meta-regression functions for 3 different types of ecosystem services able to determine the ecosystem service value for 12 different types of biomes, with the possibility of these estimates being applied at the global scale. In this study, we provide the results of a meta-analysis based on the primary value estimates from the Ecosystem Service Valuation Database [23] for 3 ecosystem services (Provisioning; Regulating and maintenance; Cultural), provided by 12 main land covers (Coastal systems; Coastal wetlands; Coral reefs; Cultivated areas; Desert; Fresh water; Grasslands; Inland wetlands; Open Ocean; Temperate/Boreal forests; Tropical forests; Woodlands). In addition, complementary explanatory variables from the World Bank Data [34] and FAOSTAT [35] were gathered. Based on this review and meta-analysis, we aim to provide recommendations for future research that may enhance the use of ecosystem service valuation for policy analysis.

The remainder of this paper is structured as follows. The "Materials and Methods" section details the MA application and use in ES studies, the theoretical specification and validation method and, finally, the ESVD database and other variables used to build the models. In turn, in the "Results" section, we expose and analyze the functional forms of the models for the three ecosystem services, present the application of the models, and discusses the results. Finally, in the "Conclusions and recommendations" section, concluding remarks and observations are presented.

2. Materials and Methods

2.1. Literature on Meta-Analysis

Benefit transfer (BT) is an economic valuation tool, with the goal to adapt value estimates from past research to assess the value of a similar, but separate, change in a different resource [36]. Technically, BT uses valuation estimates from other areas (study sites) and applies them to a similar location (policy site) [3]. It is a technique that relies on primary studies and, therefore, allows for the reduction of field research constraints, both in terms of time and infrastructure. However, it can lead to over/underestimated values while the accuracy of an ESV estimate is determined by the quality of the reference studies used. Thus, peer-reviewed empirical studies from similar biophysical and socioeconomic contexts are preferred over any other type of data source [32].

BT is useful when the estimation of the economic service value cannot be obtained due to time and/or budget constraints and to, therefore, make the best possible use of the existing literature in order to evaluate the economic importance of a natural area [19]. This is possible by adopting and applying estimates from existing studies that best suit the new context, using one or more of the following BT methods: (i) benefit estimate or value transfer, which is the extrapolation of estimates from one site to another (i.e., values are directly transposed from the study site to the policy site without amendments), (ii) benefit function transfer, which is the transfer of economic functions between the sites (i.e., coefficients are used to determine the policy site values), (iii) meta-analysis, which combines the findings of independent studies related to the research topic as to summarize the body of evidence relating to a particular issue, and (iv) preference calibration, which uses existing benefit estimates derived from different methodologies and combines them to develop a theoretically consistent estimate for policy site values [37].

The meta-analysis (MA) technique can help reduce deviations in value estimates [26]. This technique was first put forward as a research synthesis method and has since been developed and applied in many fields of research, other than the area of environmental economics [38,39]. It is widely recognized that the large and increasing literature on economic valuation of ES and environmental impacts has become difficult to interpret and that there is a need for research synthesis, especially in statistical MA, to aggregate information and insights [27,40,41].

MA is by definition a quantitative analysis of statistical summary indicators reported in a series of similar empirical studies. It is a commonly used method for compiling and analyzing the data from studies towards the creation of a value function. The method synthesizes the results of multiple studies that examine the same phenomenon, through the identification of a common effect, which is then "explained" using regression techniques in a meta-regression model [40]. In the realm of environmental resource valuation, MA is commonly used in benefit transfer endeavors due to its usefulness in incorporating a structural utility framework with less strictly economic information [27,42].

2.2. Specification of the Meta-Regression

Based on consumer rationality and reasonableness, the microeconomic consumer theory is explained by two different approaches: the indifference curve approach and the utility function approach [43]. Indifference curves represent all combinations of goods and services that provide the same level of satisfaction to an individual (i.e., the same level of global utility). Implicit in an indifference curve is the marginal rate of substitution, which expresses the maximum amount of a good that one is willing to give up in exchange for one additional unit of another good, at the same level of satisfaction [43]. Utility functions represent the degree of profitability or satisfaction that we get from using goods and services, related to a measure of satisfaction relative to an economic agent. The analysis of its variation allows for explaining the behavior that results from the decisions taken by each agent to increase his/her satisfaction.

Any meta-analytic benefit transfer (MA-BT) must be based on the ecosystem service valuation theory and the utility functions theory (see Equation (1)), specific to microeconomics [42]. The general form of an MA-BT underlying the utility function is given by [24]:

$$U_i = f(P_i,\ Y_i,\ \ Q_i, Ql_i, Sub_i,\ H_i, I_i) \tag{1}$$

where U_i is the utility (satisfaction) obtained by individual i, P_i is the general price level faced by individual i, Y_i is the individual revenue, Q_i is the quantity of ES available to individual i, Ql_i is the global quality of ES available to individual i, Sub_i represents the substitutes for Q available to individual i, H_i refers to other non-income attributes of individual i, and I_i is the information available to individual i.

Resorting to this microeconomic theoretic, we organize the MA-BT utility theory into three axes: the "strong structural utility theoretic (SSUT) approach", the "weak structural utility theoretic (WSUT) approach" and the "non-structural utility theoretic (NSUT) approach" (of which they only endorse the first two) [42].

Following the microeconomics reasoning, we assume that MA-BT is based on the utility function (see Equation (2)) and opt for analyzing the WSUT. Under the WSUT, each individual may choose between two alternative environmental options—ceteris paribus, a damaged ecosystem (Q_0) and a restored ecosystem (Q_1), which will assure an equilibrium situation (the maximum utility) [7,42], represented by:

$$U_i\ (P_i,\ Y_i,\ Q_0) = U_i\ (P_i,\ Y_i, -ESV,\ Q_1) \tag{2}$$

where U_i is the utility obtained by individual i, P_i is general price level faced by individual i, Y_i is individual revenue, Q_0 quality/quantity of ES available to individual i in the absence of any payment, ESV is ecosystem service value paid by individual i, and Q_1 is the quality/quantity of ecosystem available to individual i after having paid for these ES.

Microeconomics utility theory will hold if both sides of this parity are equal. That is, an individual will stay at the same indifference curve if he/she gets the same level of satisfaction by consuming Q_0 with no payment or by consuming Q_1 and paying ESV in exchange. That is, the ESV the individual is willing to give up must be counterbalanced by an increase in Q. Thus, $Q_1 > Q_0$ after the amount has been spent.

In this study, we adopt the WSUT approach, where variables are added in the bid-function (assumed to be derived from some unidentified utility function), but keeping the flexibility to incorporate other explanatory variables into the ESV model, such as study-site characteristics, local price levels or local individual income [7]. This is the approach used in most previous MA-BT studies [7,31,44]. Our general theoretical model will focus on estimating the ESV (see Equation (3)), as a function of various explanatory variables according to the general form of the underlying conditional indirect utility function:

$$ESV = f(B_l, SQ_l, C, QQ_r,\ I_r, P_r) \tag{3}$$

where, B_l is the biome and SQ_l the quality status for the location under analysis (l), C is the continent where the study area is located, and QQ_r is the quality/quantity of protected areas, I_r is the income and P_r is the population density in the region (r) where the study area is located.

The meta-modeling approach has several advantages for BT as compared to other methods (such as value transfer or function transfer). Different from those, which are based on single studies, MA resorts to information from a collection of studies and, thus, provides more rigorous measures of central tendencies that are sensitive to the distributions of underlying study values [24].

2.3. Validity and Reliability of a Meta-Analytic Benefit Transfer

The validity and reliability of the MA-BT can be assessed by applying the concept of transfer error (TE), defined as [32]:

$$TE = \frac{|ESV_P - ESV_B|}{ESV_B} \tag{4}$$

where ESV_P is the predicted value from the study site (s) and ESV_B is the base value ("benchmark") at the policy site. The TE is often used as a validity measure of the acceptability of meta-models. Traditionally, validity requires that the values, or the value functions generated from the study site, be statistically identical to those estimated at the policy site [8]. The main objective is to find a target value of TE = 0, confirming that the estimated values from the MA-BT values are similar to those arising from the database.

There is no agreement on maximum TE levels for BT being reliable for different policy applications. The TE analysis is not supposed to judge which levels should be considered acceptable, or even conduct traditional statistical tests of BT validity. Instead, it remains a measure of reliability, especially if TE estimates are compared across meta-model specifications and restrictions, and between alternative ways of conducting BT based on the same data [7].

Therefore, we perform the following comparisons between the estimates from the meta-model and the original observations from the database:

(a) "Value transfer" compares each ESV estimate in the database with the corresponding global mean ESV;
(b) "Global meta function transfer" compares each ESV estimate in the database with the estimates produced by the meta-model, using mean global values for the explanatory variables;
(c) "Local meta function transfer" compares each ESV estimate in the database with the estimates produced by the meta-model, using mean national values for the explanatory variables.

2.4. Background and Data

MA in environmental valuation is, generally, based on brief statistics and analytical conclusions taking a group of studies as data. Therefore, MA estimates can reduce the time spent to acquire data—both in the case of older studies and unpublished work (where data may not be available) and current studies (where authors may be slow to disclose data). However, even within the same methodology, combining primary data is not always possible due to conflicting data structures and different estimation procedures [42]. This might limit the MA studies' representativeness.

A solution to this problem is the use of specialized ESV databases, which offer a wide range of detailed information about the studies taken into account, beyond the results found in the assessment. These databases give information on other factors crucial for the delimitation of a MA model, such as: the year of the study, protection status, location, type of environment, and method. In this analysis, we use the Ecosystem Service Valuation Database (ESVD, [23]), one of the biggest databases containing real values for a range of ES and biomes where the value estimates are systematized in monetary units (€/ha/year).

The ESVD was built to process and analyze the monetary estimates of ES values from different biomes in a way that is easily used by various end-users, worldwide. Composed by 267 studies and 1310 value estimates, the ESVD links various types of information from different studies with the value estimates and case study sites. These value estimates are organized by biome, ecosystem service, and country. The main biomes are "Coastal System" (*CSys*), "Coastal Wetland" (*CWet*), "Coral Reef" (*CoRf*), "Cultivated Area" (*CuAr*), "Desert" (*Dser*), "Grasslands" (*Gras*), "Inland Wetland" (*InWt*), "Marine" (*Mari*), "Temperate or Boreal forests" (*TeFo*), "Tropical forests" (*TrFo*), "Fresh water" (*FrWa*) and "Woodland" (*Wood*). The ecosystem services are Provisioning; Regulating & maintenance and; Cultural services, divided into 14 types of services (see in Figure 1). Finally, a total of 80 countries are included, 217 values from Africa; 352 values from Asia, 208 values from Europe, 180 values from Latin America and the Caribbean; 122 values from North America, 116 from Oceania, and 114 from the whole world.

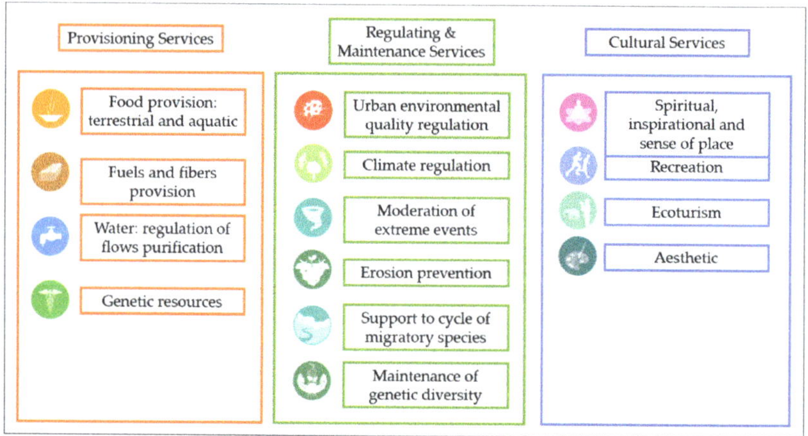

Figure 1. Ecosystem service values division [45].

Initial criteria for selecting studies from the general ESVD database were: (1) original nature of the case study data (i.e., not based on value transfer or total ecosystem value); (2) the provision of a complete set of information, including the study site location, surface area and the scale of the study (i.e., not based on a "world" scale location); (3) clear characterization of valuation methodologies used (i.e., not unknown valuation methods); (4) clear mentioning of the surface area for which the ecosystem service valuation study is applied (so that estimates of monetary values per hectare can be obtained); and (5) ES or sub-service monetary value directly linked to a specific biome/ ecosystem and unit (i.e., not per person or household). Besides information on the location of each case study, the ESVD includes information on protection status and the size of the research area, enabling for the verification of whether more estimates about the same case study location are available from other sources or publications. Together with supplementary variables, coming from complementary socio-economic databases that are added to

ESVD, these variables allow for further socio-economic interpretation of the monetary output values.

In order to relate an estimate of an ecosystem service to the socio-economic context of a case study site, two additional variables were included in the country table—namely the Gross national income (*GNI*) per capita (based on purchasing power parity in current international prices) and the average Population density (*PDen*; people per square kilometer). This information was obtained from the World Bank Data, which provides world development indicators by country [34]. Collected values were obtained for the years in which the studies were carried out.

Regarding protection status, many of the data points in the ESVD pertain to case studies in protected areas. This information allows the assessment of the influence of the protection status on ES value, testing whether protection excludes the user's access to the site and consequently to the services generated or, alternatively, whether it allows for ecosystem conservation and subsequent appreciation of the services. Protection status is classified into 3 categories: Fully protected (*FProt*), Partially protected (*PProt*) or Not protected (*NProt*). Other complementary variables collected from the World Bank Data, used to verify the study-site protection status, were: Terrestrial Protected Areas (*TProt*; the percentage of protected land by country) and Marine Protected Areas (*MProt*; the percentage of protected territorial waters). From the Food and Agriculture Organization (FAO) statistical database [35], information on the land use characteristics was collected. Namely, the percentage of forest area (*FPer*) and the percentage of agricultural land (*APer*), which helped to understand land use and occupation characteristics with emphasis on agricultural activities and state of preservation/conservation of nature.

For each biome in the ESVD, 14 ecosystem services were identified and classified into the 3 main classes: Provisioning, Regulating and maintenance, and Cultural services (see Figure 1). This classification constitutes an important step in the linkage between ES and human well-being and will be used as a basis to perform MA-BT for ecosystem valuation. Provisioning services (ESV_{Prov}) are mainly composed of food provision, water provision (including regulation of water flows and water purification), fuels and fibers provision, and genetic resources provision [45]. This is an ES highly valued by humans, because of the direct impact on our day-to-day life. Regulating and maintenance services ($ESV_{Reg\&Main}$) help maintaining air, climate, and water quality, moderating extreme events, maintaining soil quality, and preventing erosion. Albeit usually invisible and taken for granted, they are important for human well-being and the conservation of plants and animals [45]. Finally, cultural services (ESV_{Cult}) entail non-material benefits that people obtain from an ecosystem, such as aesthetic inspiration, recreation, and tourism as well as spiritual experience related to a natural environment [45].

All monetary values in the ESVD values are converted into a common reference unit, specifically 2015 'International' €/ha/year, using the Purchasing Power Parity (PPP) units expressed in Euros [34,45].

3. Results

3.1. Data Summary

Based on the above-mentioned criteria, the total number of monetary value estimates included in our sample amount to 636 observations. In this study, ES value functions are estimated for Provisioning, Regulating and maintenance, and Cultural services (see Section 3.2). The estimation of each ES value function draws on a different number of observations (see Table 1): Provisioning services (302; 47.5%), Regulating and maintenance services (225; 35.4%), and Cultural services (109; 17.1%).

Table 2 lists and describes the main variables used in the MA. Table 3 provides summary statistics for each of these variables for every service, with the exception of the dummy variables.

Table 1. Number of valuation studies, by service and biome, from the ESVD included.

Service [1]/Biome [2]	*CSys*	*CWet*	*CoRf*	*CuAr*	*Dser*	*FrWa*	*Gras*	*InWt*	*Mari*	*TeFo*	*TrFo*	*Wood*	**Total**
$\mathbf{ESV_{Prov}}$	18	55	37	6	2	5	10	75	6	8	63	17	302
$\mathbf{ESV_{Reg\&Main}}$	6	58	26	7	-	1	9	36	4	16	51	11	225
$\mathbf{ESV_{Cult}}$	7	14	42	-	-	4	2	11	4	10	14	1	109

Note: [1] ESV_{Prov} = Provisioning Ecosystem Service Values; $ESV_{Reg\&Main}$ = Regulating and maintenance Ecosystem Service Values; ESV_{Cult} = Cultural Ecosystem Service Values; [2] *CSys* = Coastal systems; *CWet* = Coastal wetlands; *CoRf* = Coral reefs; *CuAr* = Cultivated areas; *Dser* = Desert; *FrWa* = Fresh water; *Gras* = Grasslands; *InWt* = Inland wetlands; *Mari* = Marine; *TeFo* = Temp./Bor. forests; *TrFo* = Tropical forests; *Wood* = Woodland.

The common variables in all models (Provisioning; Regulating and maintenance; Cultural) are Population density (*PDen*) and Gross national income per capita (*GNI* see Table 3). These variables show the largest mean, minimum, and maximum dispersion, representing the large differences in population and wealth in countries around the world. Additional variables were created to describe potentially influential study site characteristics. In the case of Provisioning services, these were: the agricultural areas (*APer*) and the terrestrial protected areas (*TProt*). The former represents the food, fuels, and fibers provisioned, and the latter represents regulation of flows and purification provided. In the case of Regulating and maintenance services, these were: the forest areas (*FPer*) and the terrestrial (*TProt*) and marine (*MProt*) protected areas. These variables express the quality/quantity of natural resources that directly influence their prevention, moderation, and support. In the case of Cultural services, these were the marine protected areas (*MProt*), which represent quality, namely related to the sea.

Table 2. Meta-analysis variables description.

Variables	Description
APer	Agricultural land refers to the share of land area that is arable, under permanent crops, and under permanent pastures, by the percentage of land area.
FPer	Forest area with natural or planted stands of trees of at least 5 m in situ, by the percentage of land area.
MProt	Percentage of marine protected areas, from territorial waters of a country.
TProt	Percentage of terrestrial areas totally/partially protected, designated by national authorities.
GNI	Gross National Income per capita, using purchasing power parity rates.
PDen	Population density is midyear population divided by land area in square kilometers.
	Dummies
CSys; CWet; CoRf; CuAr; Dser; FrWa; Gras; InWt; Mari; TeFo; TrFo; Wood	**Biomes:** Coastal systems; Coastal wetlands; Coral reefs; Cultivated areas; Desert; Fresh water; Grasslands; Inland wetlands; Marine; Temp./Bor. forests; Tropical forests; Woodland.
Euro; Asia; Ocea; LaAm; NoAm; Afric	**Continents:** Europe; Asia; Oceania; Latin America and Caribbean; North America; Africa.
FProt; PProt; NProt	**Protection Status:** Fully protected; Partially protected; Not protected.

Table 3. Summary statistics for meta-regression variables in ecosystem services.

Variables [1]	Mean	Stand. Dev.	Min	Max
		Provisioning Services		
TProt	14.94	8.53	0.00	6.27
APer	42.78	18.99	6.27	80.89
PDen	124.60	143.77	1.70	1130.40
GNI	7481.06	10,021.30	430.00	44,740.00

Table 3. *Cont.*

		Regulating and Maintenance Services		
FPer	35.20	20.57	0.24	91.34
MProt	12.54	17.12	0.00	74.70
TProt	14.73	7.56	0.00	36.84
PDen	115.70	127.07	2.40	502.30
GNI	14,471.44	14,036.35	430.00	48,420.00
		Cultural Services		
MProt	15.52	17.63	0.00	74.82
PDen	105.23	116.90	2.30	478.30
GNI	16,750.05	13,484.39	840.00	48,420.00

Note: [1] See Table 2 for variable descriptions.

3.2. Meta-Regression Model Specification

We adopt a semi-log functional form specification for the ES value functions, which implies that the marginal effect of a change in ESV depends on income and population density [15].

The Provisioning ES value function is determined by the type of biome (D_{Biome}), location of the continent ($D_{Continent}$), terrestrial protected area (*TProt*) [46], percentage of agricultural land (*APer*) [47], population density (*PDen*) [5] and income (*GNI*) [15], and given by:

$$\ln(ESV_{Prov}) = \alpha_0 + \alpha_1 * D_{Biome} + \alpha_2 * D_{Continet} + \alpha_3 * TProt + \alpha_4 * APer + \alpha_5 * \ln(PDen) + \alpha_6 \times \ln(GNI) \quad (5)$$

where α_0 is a constant, α_1 and α_2 are dummy regression estimates, and α_3 to α_6 are variable regression estimates.

The Regulating and maintenance ES value function is determined by the type of biome (D_{biome}), location of the continent ($D_{Continent}$), level of protection in study area (*FProt*) [15], the terrestrial (*TProt*) and marine (*MProt*) protected area [46], percentage of forest land (*FPer*) [47], population density (*PDen*) [5] and income (*GNI*) [15], and given by:

$$\ln(ESV_{Reg\&Main}) = \beta_0 + \beta_1 * D_{Biome} + \beta_2 * D_{Continet} + \beta_3 * FProt + \beta_4 * FPer + \beta_5 * MProt + \beta_6 * TProt + \beta_7 * \ln(PDen) + \beta_8 * \ln(GNI) \quad (6)$$

where β_0 is a constant, β_1 and β_2 are dummy regression estimates, and β_3 to β_8 are variable regression estimates.

Finally, the Cultural ES value function is determined by the type of biome (D_{biome}), location of the continent ($D_{Continent}$), level of protection in study area (*PProt*) [15], marine protected area (*MProt*) [46], population density (*PDen*) [5] and income (*GNI*) [15], and given by:

$$\ln(ESV_{Cult}) = \gamma_0 + \gamma_1 * D_{Biome} + \gamma_2 * D_{Continet} + \gamma_3 * \mathrm{PProt} + \gamma_4 * MProt + \gamma_5 * \ln(PDen) + \gamma_6 * \ln(GNI) \quad (7)$$

where γ_0 is a constant, γ_1 and γ_2 are dummy regression estimates, and γ_3 to γ_6 are variable regression estimates.

3.3. Meta-Regression Model Results

Table 4 reports regression results for two model specifications; the "Full" model in which all variables are included and the "Restricted" model in which non-significant explanatory variables were excluded in a stepwise procedure (applying a cut-off significance level of 20% for the *t*-test). The following base values for the dummies are considered: Grasslands (*Gras*) for biomes; Not protected (*NProt*) for protection status; and Europe (*Euro*) for continents.

The main explanatory variables presented in all "Restricted" models were Population density (*ln_PDen*) and Gross national income (*ln_GNI*), with positive coefficient values and high significance (*t*-test < 0.09), which implies that an increase in population or income results in an increase ESV. As we adopt the logarithmic form for these variables, the marginal increase in ESV is decreasing in population or income.

We adopted additional explanatory variables for environmental quality, being *MProt* and *TProt* the percentage of, respectively, terrestrial and marine protected areas. Specifically, for the Provisioning model the *APer* (percentage of agricultural land) and for the Regulating & maintenance model the *FPer* (percentage of forest land) were used.

The Provisioning ES model provides a reasonable fit to the data, although it is the model with the smallest R^2 (0.19) and with the statistics of 0.01 in ANOVA for the restricted model. The signs of the explanatory variables are, as expected, positive for D_{biome}, *LaAm*, *ln_PDen*, and *ln_GNI*, and negative for *TProt*. This confirms that the other land covers analyzed tend to have a higher value than Grasslands (used as a base for the dummy biomes) and that areas located in Latin America generate larger provisioning ecosystem service values, while the ecosystem service value decreases with an increase in the percentage of protected terrestrial area. The variable *APer* is an exception (*Coef* = −0.04 and *t*-test < 0.01), presenting a negative coefficient, for which a positive sign was expected—which could be explained by the fact that countries with larger agricultural areas present a greater supply of provisioning services, though lower productivity levels. Significant explanatory variables present *t*-test < 0.19, the remaining variables were dropped. Evaluating the dummy variables for biomes, the one that presented the highest coefficient for the ESV_{Prov} was *CuAr* (*Coef* = 3.69 and *t*-test < 0.01), indicating that Cultivated areas is the key variable explaining provisioning service values.

The Regulating and maintenance ES model provides a good fit to the data, being the model with the highest R^2 (0.46) and with the statistics of 0.01 in ANOVA for the restricted model. The sign of the explanatory variables is as expected positive for D_{biome}, *ln_PDen*, and *ln_GNI*, and negative for *AFric*. This confirms that, as mentioned before, the other land covers analyzed tend to have a higher value than Grasslands and that areas located in Africa tend to have a lower value for this type of service (due to the lower aggregate income). The variables related to nature protection: *FProt* (*Coef* = −1.73 and *t*-test < 0.01), *FPer* (*Coef* = −0.02 and *t*-test < 0.05), *MProt* (*Coef* = −0.02 and *t*-test < 0.19) and *TProt* (*Coef* = −0.05 and *t*-test < 0.05), present negative coefficients, for which a positive sign was expected, revealing the theory that protected areas, which generally have low population density or are even inaccessible to the population, represent a low monetary value (i.e., people do not fully perceive the value of this service being generated). Significant explanatory variables present *t*-test < 0.19, the remaining variables were dropped. In the $ESV_{Reg\&Main}$ the largest coefficient for biome was observed in *InWt* (*Coef* = 4.77 and *t*-test < 0.01), although many others such as *CoRf*, *CWet*, and *CSys*, (*Coef* = 4.68; 4.19; 3.98 and *t*-test < 0.01, respectively) also presented high values, these biomes hold a series of important services, such as climate moderation, erosion prevention, maintenance,
and support for different species.

The Cultural ES model also presents a good fit to the data, with an R^2 (0.38) and with the statistics of 0.01 in ANOVA for the restricted model. The sign of the explanatory variables is as expected positive for *PProt*, *ln_PDen*, and *ln_GNI*, *LaAm* and negative for *MProt*, *Asia*, *Ocea*. This explains that partially protected areas make it possible for people to access and benefit from the services generated. Moreover, Latin America is the area that presents the largest Cultural ES (primary studies mainly from the Caribbean coast). The D_{biome} variables *Mari* (*Coef* = −2.47 and *t*-test < 0.12) and *TeFo* (*Coef* = −3.09 and *t*-test < 0.01), present negative coefficients, for which a positive sign was expected, due to the small number of studies related to cultural services involving these land covers in the ESVD. In the ESV_{Cult} the largest coefficient was *CoRf* (*Coef* = 2.48 and *t*-test < 0.01), explaining the high value of services associated with the Coral reefs biome, which provides services such as ecotourism, recreation, and aesthetics, receiving thousands of tourists annually.

Table 4. Meta-regression results for Provisioning (ESV_{Prov}), Regulating and maintenance ($ESV_{Reg\&Main}$) and Cultural (ESV_{Cult}) ecosystem service values.

Explanatory Variables [1]	Model Specification											
	Provisioning Serv. Model				Regu. and Main. Serv. Model				Cultural Serv. Model			
	Full		Restricted		Full		Restricted		Full		Restricted	
	Coef	*t*-Test (*sig*)	*Coef*	*t*-Test (*sig*)	*Coef*	*t*-Test (*sig*)	*Coef*	*t*-Test (*sig*)	*Coef*	*t*-Test (*sig*)	*Coef*	*t*-Test (*sig*)
CONSTANT	−3.80	0.36	−6.41	0.01	−7.97	0.03	−3.46	0.19	−12.37	0.07	−7.37	0.03
CSy	1.93	0.19	2.68	0.01	5.10	0.01	3.98	0.01	2.09	0.40	-	-
CWet	1.51	0.24	2.22	0.01	5.31	0.01	4.19	0.01	4.70	0.05	1.35	0.20
CoRf	−0.85	0.53	-	-	5.28	0.01	4.68	0.01	5.83	0.01	2.48	0.01
CuAr	3.07	0.11	3.69	0.01	4.28	0.01	3.07	0.01				
Dser	1.24	0.66	-	-								
FrWa	1.41	0.49	2.17	0.19	2.79	0.28	-	-	708	0.00	-	-
Gras [2]	-	-	-	-	-	-	-	-	-	-	-	-
InWt	1.29	0.31	2.03	0.01	5.53	0.01	4.77	0.01	5.04	0.04	1.48	0.20
Mari	1.08	0.57	2.18	0.15	2.07	0.17	-	-	1.44	0.58	−2.47	0.12
TeFo	−1.46	0.42	-	-	4.80	0.01	3.35	0.01	0.81	0.75	−3.09	0.01
TrFo	1.37	0.29	2.06	0.01	3.47	0.01	2.40	0.01	4.80	0.04	1.20	0.20
Wood	−0.33	0.83	-	-	1.69	0.12	-	-	6.28	0.08	-	-
FProt	−0.18	0.80	-	-	−1.83	0.01	−1.73	0.01	−0.42	0.75	-	-
PProt	−0.25	0.66	-	-	−0.24	0.63	-	-	0.78	0.53	1.17	0.05
NProt [2]	-	-	-	-	-	-	-	-	-	-	-	-
Euro [2]	-	-	-	-	-	-	-	-	-	-	-	-
Asia	−1.05	0.44	-	-	0.43	0.59	-	-	−1.09	0.49	−1.75	0.06
Ocea	−0.77	0.60	-	-	1.53	0.13	-	-	−0.81	0.59	−1.33	0.16
LaAm	0.82	0.55	1.76	0.01	1.14	0.23	-	-	2.55	0.17	1.33	0.18
NoAm	−0.97	0.53	-	-	0.66	0.41	-	-	0.98	0.46	-	-
Afric	−1.20	0.45	-	-	−0.79	0.49	−2.12	0.01	0.66	0.74	-	-
APer	−0.04	0.02	−0.04	0.01	-	-	-	-	-	-	-	-
FPer	-	-	-	-	−0.01	0.24	−0.02	0.05	-	-	-	-
Mprot	−0.02	0.33	-	-	−0.02	0.25	−0.02	0.19	−0.06	0.01	−0.05	0.01
TProt	−0.05	0.14	−0.05	0.10	−0.04	0.15	−0.05	0.06	−0.03	0.40	-	-
ln_*GNI*	0.81	0.01	0.87	0.01	0.65	0.02	0.49	0.03	1.21	0.02	1.04	0.01
ln_*PDen*	0.54	0.03	0.59	0.01	0.91	0.01	0.66	0.01	0.53	0.12	0.48	0.09
N	302				225				109			
R^2	0.20		0.19		0.47		0.46		0.48		0.38	
p-Value in ANOVA [3]	0.01		0.01		0.01		0.01		0.01		0.01	

Notes: Dependent variable is ln_ESV_i. [1] See Table 2 for variable descriptions; [2] Variable used as the basis for analysis of the dummies; [3] F-test of joint restriction that coefficients of excluded variables are equal to zero.

The model with the least good fit was the Provisioning ES model (R^2 = 0.19), followed by the Cultural ES model with a reasonable fit (R^2 = 0.38) and the Regulating and maintenance ES model" with a reasonably good fit (R^2 = 0.46) for the restricted models. Although these values are low as compared to other ESV meta-analysis studies (see Table 5), a great variability is observed in these studies, with R^2 between 0.25 and 0.87. The explanation for these values is related to the large number of observed studies that presented different characteristics like the location, valuation method, and different years in which the study was performed. For example, [24,30,31] presented large samples, with 682, 416, and 311 observations, respectively. In addition, these studies were applied in wide areas, covering several countries.

Table 5. Studies applying the meta-analysis for ESV.

Authors	Location	Ecosystem Service	Biome	R^2	Samp. Size	Cut-Off in *t*-Test [1]
Rosenberger & Loomis [24]	United States and Canada	Outdoor activities	-	0.26	682	0.20
Bateman & Jones [41]	British Forest—Great Britain	Recreation	Woodlands	0.71	77	0.38
Van Houtven et al. [15]	United States	Water quality	-	0.59–0.61	131	0.10
Lindhjem & Navrud [7]	Norway, Sweden, and Finland	Non-use values related to biodiversity	Forests	0.81–0.87	72	0.20
Ghermandi et al. [30]	World	Flood protection, water quality, and water storage and supply	Wetlands	0.49–0.46	416	0.10
Hjerpe et al. [29]	United States	Forest and freshwater restoration	Forests and Fresh waters	0.58–0.60	127	0.18
Rao et al. [44]	World coastal area	Shoreline protection	Coastal Areas	0.44–0.45	90	0.10
Hynes et al. [31]	World	Recreation services	Coastal Areas	0.25–0.65	311	0.10

Note: [1] Values presented for the final/best model presented.

As previously exposed, the cut-off for the significance level adopted in the t-student test for the model variables was 20%, which eventually diminished the reliability of the models (i.e., it is common to use "cut-off points" of 0.5%, 1%, 5% or even 10%). Nevertheless, authors such as [7,24,29,41] used *t*-values similar to those adopted in our research. It will be demonstrated, in the next section, that the transfer errors obtained using these value functions are smaller than those obtained using other benefit transfer techniques.

3.4. Value Function Transfer Errors and Estimates

The validity of environmental benefit transfer has been the subject of a number of studies [7,48,49]. In all of them, the validity has been tested by stating a null hypothesis of no difference between the original study result and the benefit transfer estimate [50]. As in those studies, in this study we seek to verify the differences between the estimated values from MA-BT with the values from the ESVD database, using the Transfer Error technique (see Section 2.3).

3.4.1. Transfer Errors

To assess the accuracy of the estimated ES value meta-models, in order to justify their adoption in future research covering different locations with varied characteristics, we determined the transfer errors associated with Value transfer, Global meta function transfer, and Local meta function transfer (see Section 2.3). This is done for the Provisioning, Regulating and maintenance, and Cultural ES value functions (see Tables 6–8, respectively).

The ecosystem service values and transfer errors per biome related to the estimates for the Provisioning ES are presented in Table 6. Overall, it can be concluded that the transfer error is reduced when moving from Value transfer to Global meta function transfer and, in turn, that the transfer error is further reduced when moving to Local meta function transfer. Notable exception holds for *Wood*, which demonstrates the lowest transfer error when using Value transfer. This is explained by the fact that this variable was dropped from the restricted model (not significant according to the *t*-test). Also, in some cases, the transfer error increases slightly when moving from Global meta function transfer to Local meta function transfer (such as for *FrWa*, *Mari*, and *TeFo*), which is explained by the large variation of values in the ESVD database that contained studies from different countries, continents, and years, and in the case of those biomes, ranging from 1.5 to 3000.0 €/ha/year.

Table 6. Comparison of values and transfer errors (TE) per biome for Provisioning ES, based on Value transfer, Global meta function transfer, and Local meta function transfer (in 2015 €/ha/yr).

	Value Transfer		Global Meta Function Transfer		Local Meta Function Transfer	
Biome [1]	**Value**	**TE (ETE1)**	**Value**	**TE (ETE2)**	**Value**	**TE (ETE3)**
CSys	1336.0	926.2	81.9	56.7	185.7	11.4
CWet	362.7	1228.2	30.7	103.9	66.0	10.1
CoRf	1463.7	7.0×10^6	10.3	5.0×10^4	23.1	1.6×10^4
CuAr	2795.2	4.2×10^5	141.7	2.2×10^4	741.8	1.4×10^4
Dser	82.5	106.2	1.5	2.0	1.5	2.0
FrWa	594.9	107.3	59.7	10.7	120.5	15.6
Gras	164.9	4.5×10^4	2.8	769.9	8.1	106.3
InWt	176.8	2013.6	6.2	71.0	15.8	54.7
Mari	50.8	2.76	27.6	1.4	48.0	4.0
TeFo	68.1	203.2	10.8	32.1	14.3	41.0
TrFo	277.3	297.8	31.2	33.3	58.3	19.6
Wood	110.6	1.1×10^6	4.6	2.7×10^6	15.4	6.5×10^6

Note: [1] *CSys* = Coastal systems; *CWet* = Coastal wetlands; *CoRf* = Coral reefs; *CuAr* = Cultivated areas; *Dser* = Desert; *FrWa* = Fresh water; *Gras* = Grasslands; *InWt* = Inland wetlands; *Mari* = Marine; *TeFo* = Temp./Bor. forests; *TrFo* = Tropical forests; *Wood* = Woodland.

Table 7. Comparison of values and transfer errors (TE) per biome for Regulating and maintenance ES, based on Value transfer, Global meta function transfer, and Local meta function transfer (in 2015 €/ha/yr).

	Value Transfer		Global Meta Function Transfer		Local Meta Function Transfer	
Biome [1]	**Value**	**TE (ETE1)**	**Value**	**TE (ETE2)**	**Value**	**TE (ETE3)**
CSys	941.9	7.6	258.3	1.8	1381.8	3.8
CWet	5088.3	267.3	430.6	22.5	943.2	12.3
CoRf	7071.0	3189.4	383.9	173.0	1236.6	18.6
CuAr	425.6	20.0	134.4	6.2	215.0	1.7
Dser	-	-	-	-	-	-
FrWa	115.5	0.0	29.8	0.7	29.8	0.7
Gras	111.9	1464.9	11.7	153.3	22.2	37.2
InWt	1660.2	1430.5	188.8	162.6	747.4	17.6
Mari	748.3	260.8	18.0	6.2	28.7	1.4
TeFo	641.8	44.9	94.7	6.4	197.2	5.4
TrFo	135.7	111.0	16.4	13.1	48.4	9.6
Wood	199.0	117.5	17.9	10.7	41.4	25.0

Note: [1] See Table 6 for variable descriptions.

Table 7 presents the ecosystem service values and transfer errors per biome associated with the estimates for the Regulating and maintenance ES. According to the analysis of the previous table, the TE is reduced when moving from Value transfer to Global meta function transfer and then moving to Local meta function transfer. In this case, the exceptions hold for *CSys* and *Wood*, which demonstrate the lowest transfer error when using Global meta function transfer. This is explained by the variation of the values presented in the ESVD database for these biomes. No transfer error is observed for *FrWa* when using value transfer, as only one observation for this biome is available in the ESVD. Finally, no value estimate and transfer error were calculated for *Dser* because there are no primary value estimates data for this biome in the ESVD.

Table 8. Comparison of values and transfer errors (TE) per biome for Cultural ES, based on Value transfer, Global meta function transfer, and Local meta function transfer (in 2015 €/ha/yr).

Biome [1]	Value Transfer		Global Meta Function Transfer		Local Meta Function Transfer	
	Value	**TE (ETE1)**	**Value**	**TE (ETE2)**	**Value**	**TE (ETE3)**
CSys	156.9	156.3	90.6	90.0	186.9	33.7
CWet	3099.8	119.3	152.6	5.6	267.0	5.1
CoRf	5340.9	2 138.0	308.9	123.6	1695.3	17.1
CuAr	-	-	-	-	-	-
Dser	-	-	-	-	-	-
FrWa	651.4	0.5	16.1	1.0	36.2	0.9
Gras	1.4	0.2	48.6	35.3	58.2	46.4
InWt	681.5	15.3	142.4	3.0	234.0	3.3
Mari	311.8	316.9	7.4	7.2	20.6	1.6
TeFo	878.8	1.9×10^4	9.1	204.8	13.2	180.5
TrFo	275.4	38.3	38.0	5.1	85.6	6.2
Wood	3840.5	0.0	196.7	0.9	196.7	0.9

Note: [1] See Table 6 for variable descriptions.

Finally, Table 8 presents the ecosystem service values and transfer errors per biome associated with the estimates for the Cultural ES. Again, it can be observed that the transfer error is reduced when moving from Value transfer to Global meta function transfer and next, to Local meta function transfer. Although there are exceptions, such as for *FrWa*, *InWt*, and *TrFo*, which presented similar TE across Global and Local meta function transfer. One prominent exception holds for *Gras*, which demonstrates the lowest transfer error when using Value transfer. This is justified because it contained only two observations for this biome in the database. No transfer error is observed for *Wood* when using value transfer, as only one observation for this biome is available in the ESVD. Finally, no value estimates and transfer errors were calculated for *CuAr* and *Dser* because there are no primary value estimates for these biomes in the ESVD.

Hence, it can be concluded that transfer errors are reduced significantly when using Global meta function transfer and, in particular, Local meta function transfer as compared to Value transfer. This is justified because value function transfers allow the analyst greater control over differences across sites, they can yield lower transfer errors than simple mean value transfers [51]. In fact, by comparison, value functions offer a greater reflection of the variability of a sample, because the study is dealing with a database with great variability. For this reason, finding a model that, for the most part, has obtained a superior result than other benefit transfers techniques, is an advance that justifies its application given the heterogeneity of the data.

Value functions should thereby draw upon common drivers of preferences reflected in economic theory, including only those variables applicable to all sites [52]. Economic theory suggests that the benefits from environmental improvements should be determined by [53]: (i) change in provision, (ii) distance to the site, (iii) distance to substitute sites, and (iv) characteristics of the valuing individual (in particular income). That is why

Local meta function transfer presents the lowest TE, for addressing these preferences and reflecting the context of each country.

3.4.2. Local Value Function Transfer Estimates

Ecosystem service value estimates per biome for Provisioning (ESV_{Prov}), Regulating and maintenance ($ESV_{Reg\&Main}$), and Cultural (ESV_{Cult}) ecosystem services, are presented in Table 9. Value estimates are thereby based on the restricted models presented in Table 4, using local value function transfer and mean values for the explanatory variables (from Table 3).

Table 9. Estimated ES values per biome for Provisioning (ESV_{Prov}), Regulating and maintenance ($ESV_{Reg\&Main}$), and Cultural (ESV_{Cult}) ecosystem services, using Local meta function transfer and mean national values for the explanatory variables (in 2015 €/ha/yr).

Ecosystem Service [1]	*CSys*	*CWet*	*CoRf*	*CuAr*	*Dser*	*FrWa*	*Gras*	*InWt*	*Mari*	*TeFo*	*TrFo*	*Wood*
ESV_{Prov}	44.5	28.0	3.0	122.0	3.0	26.7	3.0	23.1	27.0	3.0	23.9	3.0
$ESV_{Reg\&Main}$	193.2	238.1	389.9	78.1	-	3.6	3.6	425.8	3.6	103.3	39.9	3.6
ESV_{Cult}	127.1	491.6	1520.7	-	-	127.1	127.1	555.3	10.8	5.8	420.8	127.1
ESV_{Total}	364.8	757.7	1913.6	200.1	3.0	157.4	133.7	1004.2	41.3	112.2	484.5	133.7

Note: [1] ESV_{Prov} = Provisioning Ecosystem Service Values; $ESV_{Reg\&Main}$ = Regulating and maintenance Ecosystem Service Values; ESV_{Cult} = Cultural Ecosystem Service Values; ESV_{Total} = Total Ecosystem Service Values.

The values found in Table 9 show great variability, with values ranging from ESV_{Total} = 3.0 €/ha/year for Desert areas to ESV_{Total} = 1913.5 €/ha/year for Coral reefs. The biomes that provide largest total economic value are Coral reefs (*CoRf* = 1913.6 €/ha/year), Inland wetlands (*InWt* = 1004.2 €/ha/year) and Coastal wetlands (*CWet* = 757.7 €/ha/year). These biomes, in addition to standing out for providing a great diversity of ecosystem services, are also the smallest biomes in terms of the area around the globe and, consequently, the scarcest and, thus, most valuable. In fact, in studies that analyzed ES globally [4,5,54], these biomes were also those with the highest value.

Provisioning services represent the lowest values and are related to the supplies of products (such as food, materials, or water) with values close to their direct use values [5]. The largest provisioning ES values are provided by Cultivated areas (*CuAr* = 121.9 €/ha/year) and Coastal System (*CSys* = 44.5 €/ha/year). The lowest values were found for Coral reefs, Desert, Grasslands, and Temp./Bor. forests (*CoRf, Dser, Gras,* and *TeFo*, with a value of 3.0 €/ha/year each).

Regulating and maintenance services are linked to more indirect benefits, which are related to quality, moderation, and prevention in environmental factors (Rao et al., 2015). The largest Regulating & maintenance ES values are provided by Inland wetlands (*InWt* = 425.8 €/ha/year), followed by Coral reefs (*CoRf* = 389.9 €/ha/year) and Coastal wetlands (*CWet* = 238.1 €/ha/year), demonstrating a high added value for areas in transition, notably coastal areas. The lowest values were found for the Marine and Woodland areas (*Mari* and *Wood*, with a value of 3.6 €/ha/year each).

Cultural services represent the largest values, because they involve complex issues such as aesthetics, generated inspiration, and spirituality, which can be considered incommensurable values as the perception about the environment varies from person to person [5,31]. The largest cultural ES values are provided by Coral reefs (*CoRf* = 1520.7 €/ha/year), Inland wetlands (*InWt* = 555.3 €/ha/year) and Coastal wetlands (*CSys* = 491.6 €/ha/year). The lowest values were found for the Marine areas (*Mari* = 10.8 €/ha/year) and Temp./Bor. forests (*TeFo* = 5.8 €/ha/year).

It is necessary to be cautious when valuing ecosystem services since, although the aim of pricing is to use values in monetary units, they serve as a tool to provide better insight into the economic benefits of ecosystem goods and services. We do not try to find the shortcomings and limitations of monetary valuation, both in relation to ecosystem services and man-made goods and services [5,55].

When ESV's models are created and values for biomes are estimated, this does not mean the biomes in question should be treated as private commodities that can be traded in private markets. Most of those ecosystem services are public goods or the product of common assets that cannot, or should not, be sold. Although the flowers, fruits, wood, and leaves enter the market as private goods, the ecosystems that produce them, as for example forests and woodlands, are common assets. Their values are an estimation of the benefits to society expressed in a way that communicates with a broad audience. This can help to raise awareness of the importance of ecosystem services to society and serve as a powerful and essential communication tool to inform better, more balanced decisions regarding trade-offs with policies that enhance the gross domestic product but damage ecosystem services [4].

4. Conclusions and Recommendations

Ecosystem service value (ESV) meta-models were designed to provide access to values in monetary units for ecosystem services (ES), taking into account the local context of the country and area under analysis. Through their application, it is possible to estimate values for 3 different types of ecosystem services (Provisioning; Regulating and maintenance; Cultural) and 12 different types of land covers (Coastal systems; Coastal wetlands; Coral reefs; Cultivated areas; Desert; Fresh water; Grasslands; Inland wetlands; Open ocean; Temperate/Boreal forests; Tropical forests; Woodlands) in the world. To this end, we built on the review and meta-analysis of the Ecosystem Service Valuation Database (ESVD).

The highest ES values were those associated with Cultural services, followed by Regulating and maintenance and, finally, Provisioning services. Among the biomes with greater associated ecosystem service values are Coral reefs, Inland wetlands, and Coastal wetlands that, among other characteristics, are transitional, aquatic-terrestrial biomes that are scarce and provide a great diversity of services.

It was observed that local independent variables, such as income, population, agricultural and forest area, and those related to the level of environmental protection, are significant explanatory variables and, thus, comprise the ESV meta-models. The application of the meta-functions provides values with greater accuracy as compared to simple value transfer and, as shown by the transfer error analysis, the application of local variables (local meta function transfer) further increases this precision.

A meta-analysis, thus, reduces value transfer errors by taking into account local specifications to determine ESV's. There are several studies that have used meta-models for the valuation of specific ecosystem services and biomes (e.g., [15,29–31]), however, we have not found such a comprehensive study in the literature that has determined the value of 3 ecosystem services for 12 different biomes in the world. Even considering that there are certain transfer errors with the application of meta-models, as compared to other benefit transfer techniques (such as value transfer and value function transfer) the meta-analysis technique has shown to be the best way to estimate the value of ecosystem services.

Some caveats to this study remain. First, there are improvements that can be made to the results, such as updating the database, adopting other explanatory variables, or even different functional forms. Second, the adoption of the ESVD, which although very broad, has some limitations, such as the necessity for further studies for biomes such as Fresh water, which presented only one study for Regulating and maintenance ES, and Woodland, which presented only one study for Cultural ES. Moreover, biomes such as Cultivated areas, Desert, and Marine presented few valuation studies, which could directly influence their estimated ES values. Third, ecosystem services and values from marine biomes face particular challenges as these are scarcely studied and poorly understood (e.g., [56,57]). Finally, it was not possible to estimate values for urban areas, albeit they are important because they have a constant relationship with human well-being through services provided by areas such as parks, squares, and green spaces, as there were no studies analyzing this land cover in the ESVD database.

We expect this study to be a step further in studies that involve valuing ecosystem services and provide a basis for future research. Not in the least because ecosystem services and values are increasingly considered in environmental planning and nature conservation. Using reliable ecosystem service value estimates from local value functions for 3 ecosystem service types across 12 biomes will facilitate this process—in particular in data-poor circumstances.

Author Contributions: Conceptualization, L.M.F., P.R. and M.I.B.; data curation, L.M.F., P.R. and W.R.; formal analysis, L.M.F., P.R. and G.O.; investigation, L.M.F., P.R. and G.O.; methodology, L.M.F. and P.R.; supervision, P.R.; validation, W.R. and P.R.; writing—original draft, L.M.F., P.R. and M.I.B.; writing—review and editing, L.M.F., P.R., M.I.B. and W.R. All authors have read and agreed to the published version of the manuscript.

Funding: This research was funded by Coordenação de Aperfeiçoamento de Pessoal de Nível Superior—Brazil (CAPES), who financed part of this work through—Finance Code 001. Thanks are also due, for the financial support, to CESAM (UIDB/50017/2020 and UIDP/50017/2020), to FCT/MCTES through national funds, and to the co-funding by European funds when applicable.

Institutional Review Board Statement: Not applicable.

Informed Consent Statement: Not applicable.

Data Availability Statement: ESVD. Ecosystem Service Valuation Database. Foundation for Sustainable Development. Available online: https://www.es-partnership.org/esvd (accessed on 18 June 2021). World Bank. The World Bank—Databank. World Bank Group. Available online: https://databank.worldbank.org/ (accessed on 18 June 2021). FAOSTAT. Food and Agriculture Data. Food and Agriculture Organization of the United Nations. Available online: http://www.fao.org/faostat/en/#data (accessed on 18 June 2021).

Acknowledgments: We would like to thank the Centre for Environmental and Marine Studies (CESAM) and the Department of Environment and Planning (DAO) at the University of Aveiro for facilitating this research. The authors also thank the anonymous referees for their helpful comments on earlier versions of this paper.

Conflicts of Interest: The authors declare no conflict of interest.

References

1. Say, J.B. *Cours Complet D'économie Politique Pratique*; Nabu Press: Paris, France, 1829; p. 790.
2. Bordt, M.; Saner, M. Which ecosystems provide which services? A meta-analysis of nine selected ecosystem services assessments. *One Ecosyst.* **2019**, *4*, e31420. [CrossRef]
3. Brouwer, R. Environmental value transfer: State of the art and future prospects. *Ecol. Econ.* **2000**, *32*, 137–152. [CrossRef]
4. Costanza, R.; De Groot, R.; Sutton, P.; Van der Ploeg, S.; Anderson, S.J.; Kubiszewski, I.; Farber, S.; Turner, R.K. Changes in the global value of ecosystem services. *Glob. Environ. Chang.* **2014**, *26*, 152–158. [CrossRef]
5. De Groot, R.; Brander, L.; Van Der Ploeg, S.; Costanza, R.; Bernard, F.; Braat, L.; Christie, M.; Crossman, N.; Ghermandi, A.; Hein, L.; et al. Global estimates of the value of ecosystems and their services in monetary units. *Ecosyst. Serv.* **2012**, *1*, 50–61. [CrossRef]
6. Díaz, S.; Demissew, S.; Carabias, J.; Joly, C.; Lonsdale, M.; Ash, N.; Larigauderie, A.; Adhikari, J.R.; Arico, S.; Báldi, A.; et al. The IPBES Conceptual Framework—Connecting nature and people. *Curr. Opin. Environ. Sustain.* **2015**, *14*, 1–16. [CrossRef]
7. Lindhjem, H.; Navrud, S. How reliable are meta-analyses for international benefit transfers? *Ecol. Econ.* **2008**, *66*, 425–435. [CrossRef]
8. Navrud, S.; Ready, R. Lessons Learned for Environmental Value Transfer. In *Environmental Value Transfer: Issues and Methods*; Navrud, S., Ready, R., Eds.; Springer: Amsterdam, The Netherlands, 2007; pp. 283–290.
9. Alcamo, J. *Ecosystems and Human Well-Being: A Framework for Assessment*; Island Press: Washington, DC, USA, 2003; p. 266.
10. Haines-Young, R.; Potschin, M.B. The Common International Classification of Ecosystem Services. European Environment Agency. Available online: https://www.cices.eu (accessed on 16 June 2021).
11. Crossman, N.D.; Bryan, B.A. Identifying cost-effective hotspots for restoring natural capital and enhancing landscape multifunctionality. *Ecol. Econ.* **2009**, *68*, 654–668. [CrossRef]
12. Crossman, N.D.; Bryan, B.A.; Summers, D.M. Carbon payments and low-cost conservation. *Conserv. Biol.* **2011**, *25*, 835–845. [CrossRef]

13. WBCSD. *Biodiversity and Ecosystem Services: Scaling Up Business Solutions: Company Case Studies That Help Achieve Global Biodiversity Targets;* WBCSD Ecosystems: Geneva, Switzerland, 2012; p. 67.
14. Maynard, S.; James, D.; Davidson, A. The development of an ecosystem services framework for South East Queensland. *Environ. Manag.* **2010**, *45*, 881–895. [CrossRef]
15. Van Houtven, G.; Powers, J.; Pattanayak, S.K. Valuing water quality improvements in the United States using meta-analysis: Is the glass half-full or half-empty for national policy analysis? *Resour. Energy Econ.* **2007**, *29*, 206–228. [CrossRef]
16. Schröter, M.; Barton, D.N.; Remme, R.P.; Hein, L. Accounting for capacity and flow of ecosystem services: A conceptual model and a case study for Telemark, Norway. *Ecol. Indic.* **2014**, *36*, 539–551. [CrossRef]
17. Perez-Verdin, G.; Sanjurjo-Rivera, E.; Galicia, L.; Hernandez-Diaz, J.C.; Hernandez-Trejo, V.; Marquez-Linares, M.A. Economic valuation of ecosystem services in Mexico: Current status and trends. *Ecol. Serv.* **2016**, *21*, 6–19. [CrossRef]
18. Schröter, M.; Albert, C.; Marques, A.; Tobon, W.; Lavorel, S.; Maes, J.; Brown, C.; Klotz, S.; Bonn, A. National Ecosystem Assessments in Europe: A Review. *BioScience* **2016**, *66*, 813–828. [CrossRef]
19. Alves, F.; Roebeling, P.; Pinto, P.; Batista, P. Valuing ecosystem service losses from coastal erosion using a benefits transfer approach: A case study for the Central Portuguese coast. *J. Coast. Res.* **2009**, *56*, 1169–1173.
20. Paprotny, D.; Terefenko, P.; Giza, A.; Czapliński, P.; Vousdoukas, M.I. Future losses of ecosystem services due to coastal erosion in Europe. *Sci. Total Environ.* **2021**, *760*, 144310. [CrossRef]
21. Roebeling, P.; Costa, L.; Magalhães-Filho, L.; Tekken, V. Ecosystem service value losses from coastal erosion in Europe: Historical trends and future projections. *J. Coast. Conserv.* **2013**, *17*, 389–395. [CrossRef]
22. De Civita, P.; Filion, F.; Frehs, J. EVRI-Environmental Valuation Reference Inventory. Environment and Climate Change Canada. Available online: https://www.evri.ca (accessed on 20 December 2020).
23. ESVD. Ecosystem Service Valuation Database. Foundation for Sustainable Development. Available online: https://www.es-partnership.org/esvd (accessed on 20 December 2020).
24. Rosenberger, R.; Loomis, J. Using meta-analysis for benefit transfer: In-sample convergent validity tests of an outdoor recreation database. *Water Res. Resour.* **2000**, *36*, 1097–1107. [CrossRef]
25. Desvousges, W.; Johnson, F.; Banzhaf, H. *Environmental Policy Analysis with Limited Information: Principles and Applications of the Transfer Method;* Edward Elgar: Northampton, UK, 1998; p. 244.
26. Hoehn, J. Methods to address selection effects in the meta regression and transfer of ecosystem values. *Ecol. Econ.* **2006**, *60*, 389–398. [CrossRef]
27. Smith, V.K.; Pattanayak, S.K. Is meta-analysis a Noah's Ark for non-market valuation? *Environ. Resour. Econ.* **2002**, *22*, 271–296. [CrossRef]
28. Johnston, R.J.; Besedin, E.Y.; Iovanna, R.; Miller, C.J.; Wardwell, R.F.; Ranson, M.H. Systematic variation in willingness to pay for aquatic resource improvements and implications for benefit transfer: A meta-analysis. *Can. J. Agric. Econ./Rev. Can. D'agroeconomie* **2005**, *53*, 221–248. [CrossRef]
29. Hjerpe, E.; Hussain, A.; Phillips, S. Valuing type and scope of ecosystem conservation: A meta-analysis. *J. For. Econ.* **2015**, *21*, 32–50. [CrossRef]
30. Ghermandi, A.; van den Bergh, J.C.J.M.; Brander, L.M.; de Groot, H.L.F.; Nunes, P.A.L.D. Values of natural and human-made wetlands: A meta-analysis. *Water Res. Resour.* **2010**, *46*, W12516. [CrossRef]
31. Hynes, S.; Ghermandi, A.; Norton, D.; Williams, H. Marine recreational ecosystem service value estimation: A meta-analysis with cultural considerations. *Ecosyst. Serv.* **2018**, *31*, 410–419. [CrossRef]
32. Ready, R.; Navrud, S. International benefit transfer: Methods and validity tests. *Ecol. Econ.* **2006**, *2*, 429–434. [CrossRef]
33. Subroy, V.; Gunawardena, A.; Polyakov, M.; Pandit, R.; Pannell, D.J. The worth of wildlife: A meta-analysis of global non-market values of threatened species. *Ecol. Econ.* **2019**, *164*, 106374. [CrossRef]
34. World Bank. The World Bank—Databank. World Bank Group. Available online: https://databank.worldbank.org/ (accessed on 20 June 2020).
35. FAOSTAT. Food and Agriculture Data. Food and Agriculture Organization of the United Nations. Available online: http://www.fao.org/faostat/en/#data (accessed on 20 December 2020).
36. Smith, V.K.; Van Houtven, G.; Pattanayak, S.K. Benefit transfer via preference calibration: 'Prudential algebra' for policy. *Land Econ.* **2002**, *78*, 132–152. [CrossRef]
37. Bergstrom, J.C.; De Civita, P. Status of benefit transfer in the United States and Canada: A review. *Can. J. Agric. Econ.* **1999**, *47*, 79–87. [CrossRef]
38. Glass, G.V. Primary, secondary and meta-analysis of research. *Educ. Res.* **1976**, *5*, 3–8. [CrossRef]
39. Nelson, J.P.; Kennedy, P.E. The use (and abuse) of meta-analysis in environmental and natural resource economics: An assessment. *Environ. Resour. Econ.* **2009**, *42*, 345–377. [CrossRef]
40. Stanley, T.D. Wheat from chaff: Meta-analysis as quantitative literature review. *J. Econ. Pers.* **2001**, *15*, 131–150. Available online: https://www.jstor.org/stable/2696560 (accessed on 20 December 2020). [CrossRef]
41. Bateman, I.J.; Jones, A.P. Contrasting conventional with multi-level modeling approaches to meta-analysis: Expectation consistency in U.K. Woodland recreation values. *Land Econ.* **2003**, *79*, 235–258. [CrossRef]
42. Bergstrom, J.C.; Taylor, L.O. Using meta-analysis for benefits transfer: Theory and practice. *Ecol. Econ.* **2006**, *60*, 351–360. [CrossRef]

43. Pindyck, R.S.; Rubinfeld, D.L. *Microeconomics*, 5th ed.; Prentice Hall: São Paulo, Brazil, 2005; p. 771.
44. Rao, N.; Ghermandi, A.; Portela, R.; Wang, X. Global values of coastal ecosystem services: A spatial economic analysis of shoreline protection. *Ecosyst. Serv.* **2015**, *11*, 95–105. [CrossRef]
45. FAO. *State of World Fisheries and Aquaculture*; Sustainability in Action; FAO: Rome, Italy, 2020; p. 196.
46. Agardy, M.T. Advances in marine conservation: The role of marine protected areas. *Trends Ecol. Evol.* **1994**, *9*, 267–270. [CrossRef]
47. De Beurs, K.M.; Henebry, G.M. Land surface phenology, climatic variation, and institutional change: Analyzing agricultural land cover change in Kazakhstan. *Remote Sens. Environ.* **2004**, *89*, 497–509. [CrossRef]
48. Brouwer, R.; Spaninks, A. The Validity of Environmental Benefit Transfer: Further Empirical Testing. *Environ. Resour. Econ.* **1999**, *14*, 95–117. [CrossRef]
49. Bergland, O.; Magnussen, K.; Navrud, S. Benefit Transfer: Testing for Accuracy and Reliability. In *Comparative Environmental Economic Assessment*; Florax, R.J.G.M., Nijkamp, P., Willis, K., Eds.; Edward Elgar: Cheltenham, UK, 2002.
50. Kristofersson, D.; Navrud, S. Validity tests of benefit transfer—Are we performing the wrong tests? *Environ. Resour. Econ.* **2005**, *30*, 279–286. [CrossRef]
51. Pearce, D.W.; Whittington, D.; Georgiou, S. *Project and Policy Appraisal: Integrating Economics and Environment*; OECD: Paris, France, 1994; p. 346.
52. Roebeling, P.; Abrantes, N.; Ribeiro, S.; Almeida, P. Estimating cultural benefits from surface water status improvements in freshwater wetland ecosystems. *Sci. Total Environ.* **2016**, *545*, 219–226. [CrossRef]
53. Bateman, I.J.; Brouwer, R.; Ferrini, S.; Schaafsma, M.; Barton, D.N.; Dubgaard, A.; Hasler, B.; Hime, S.; Liekens, I.; Navrud, S.; et al. Making benefit transfers work: Deriving and testing principles for value transfers for similar and dissimilar sites using a case study of the non-market benefits of water quality improvements across Europe. *Environ. Resour. Econ.* **2011**, *50*, 365–387. [CrossRef]
54. Costanza, R.; d'Arge, R.; De Groot, R.; Farber, S.; Grasso, M.; Hannon, B.; Limburg, K.; Naeem, S.; O'Neill, R.V.; Paruelo, J.; et al. The value of the world's ecosystem services and natural capital. *Nature* **1997**, *387*, 253–259. [CrossRef]
55. Thompson, B.H.; Paradise, R.E.; Mccarty, P.; Segerson, K.; Ascher, W.; McKenna, D.C.; Biddinger, G. *Valuing the Protection of Ecological Systems and Services. A Report of the EPA Science Advisory Committee, EPA-SAB-09-012*; Science Advisory Board, U.S.U.S. Environmental Protection Agency Science Advisory Board: Washington, DC, USA, 2009; p. 138.
56. Townsend, M.; Davies, K.; Hanley, N.; Hewitt, J.E.; Lundquist, C.J.; Lohrer, A.M. The Challenge of Implementing the Marine Ecosystem Service Concept. *Front. Mar. Sci.* **2018**, *5*, 359. [CrossRef]
57. Austen, M.C.; Andersen, P.; Armstrong, C.; Döring, R.; Hynes, S.; Levrel, H.; Oinonen, S.; Ressurreição, A. Valuing Marine Ecosystems: Taking into Account the Value of Ecosystem Benefits in the Blue Economy. In *EMB Future Science Brief. N.5*; European Marine Board: Oostende, Belgium, 2019; p. 36.

MDPI
St. Alban-Anlage 66
4052 Basel
Switzerland
Tel. +41 61 683 77 34
Fax +41 61 302 89 18
www.mdpi.com

Environments Editorial Office
E-mail: environments@mdpi.com
www.mdpi.com/journal/environments

www.ingramcontent.com/pod-product-compliance
Lightning Source LLC
LaVergne TN
LVHW070612170726
843515LV00004B/52

* 9 7 8 3 0 3 6 5 2 2 3 5 7 *